Random Graphs and Networks: A First Course

Networks surround us, from social networks to protein–protein interaction networks within the cells of our bodies. The theory of random graphs provides a necessary framework for understanding their structure and development.

This text provides an accessible introduction to this rapidly expanding subject. It covers all the basic features of random graphs – component structure, matchings and Hamilton cycles, connectivity, and chromatic number – before discussing models of real-world networks, including intersection graphs, preferential attachment graphs, and small-world models.

Based on the authors' own teaching experience, *Random Graphs and Networks: A First Course* can be used as a textbook for a one-semester course on random graphs and networks at advanced undergraduate or graduate level. The text includes numerous exercises, with a particular focus on developing students' skills in asymptotic analysis. More challenging problems are accompanied by hints or suggestions for further reading.

Alan Frieze is Professor in the Department of Mathematical Sciences at Carnegie Mellon University. He has authored almost 400 publications in top journals and was a plenary speaker at the 2014 International Congress of Mathematicians.

Michał Karoński is Professor Emeritus in the Faculty of Mathematics and Computer Science at Adam Mickiewicz University, where he founded the Discrete Mathematics group. He served as Editor-in-Chief of *Random Structures and Algorithms* for 30 years.

Random Graphs and Networks: A First Course

ALAN FRIEZE
Carnegie Mellon University

MICHAŁ KAROŃSKI
Adam Mickiewicz University

Shaftesbury Road, Cambridge CB2 8EA, United Kingdom

One Liberty Plaza, 20th Floor, New York, NY 10006, USA

477 Williamstown Road, Port Melbourne, VIC 3207, Australia

314–321, 3rd Floor, Plot 3, Splendor Forum, Jasola District Centre,
New Delhi – 110025, India

103 Penang Road, #05–06/07, Visioncrest Commercial, Singapore 238467

Cambridge University Press is part of Cambridge University Press & Assessment,
a department of the University of Cambridge.

We share the University's mission to contribute to society through the pursuit of
education, learning and research at the highest international levels of excellence.

www.cambridge.org
Information on this title: www.cambridge.org/9781009260282

DOI: 10.1017/9781009260268

First published 2023

A catalogue record for this publication is available from the British Library

A Cataloging-in-Publication data record for this book is available from the Library of Congress

ISBN 978-1-009-26028-2 Hardback
ISBN 978-1-009-26030-5 Paperback

To our grandchildren

Contents

Preface

In 2016, the Cambridge University Press published our book entitled *Introduction to Random Graphs* (see [52]). In the preface, we stated that our purpose in writing it was

> ... to provide a gentle introduction to a subject that is enjoying a surge in interest. We believe that the subject is fascinating in its own right, but the increase in interest can be attributed to several factors. One factor is the realization that networks are "everywhere". From social networks such as Facebook, the World Wide Web and the Internet to the complex interactions between proteins in the cells of our bodies, we face the challenge of understanding their structure and development. By and large, natural networks grow unpredictably, and this is often modeled by a random construction. Another factor is the realization by Computer Scientists that NP-hard problems are typically easier to solve than their worst-case suggests, and that an analysis of running times on random instances can be informative.

After five years since the completion of *Introduction to Random Graphs*, we have decided to prepare a slimmed down, reorganized version, at the same time supplemented with some new material. After having taught graduate courses on topics based on material from our original book and after having heard suggestions from our colleagues, teaching similar courses, we decided to prepare a new version which could be used as a textbook, supporting a one-semester undergraduate course for mathematics, computer science, as well as physics majors interested in random graphs and network science.

Based on our teaching experience, the goal of this book is to give our potential reader the knowledge of the basic results of the theory of random graphs and to show how it has evolved to build firm mathematical foundations for modern network theory, in particular, in the analysis of real-world networks. We have supplemented theoretical material with an extended description of the basic analytic tools used in the book, as well as with many exercises and problems. We sincerely hope that our text will encourage our potential reader to continue the study of random graphs and networks on a more advanced level in the future.

Acknowledgments

Our special thanks go to Katarzyna Rybarczyk for her careful reading of some early chapters of this book. We are particularly grateful to Mihyun Kang and her colleagues from the University of Graz for their detailed comments and useful suggestions: Tuan Anh Do, Joshua Erde, Michael Missethan, Dominik Schmid, Philipp Sprüssel, Julian Zalla.

Conventions/Notations

Often in what follows, we will give an expression for a large positive integer. It might not be obvious that the expression is actually an integer. In such cases, the reader can rest assured that he/she can round up or down and obtain any required property. We avoid this rounding for convenience and for notational purposes.

In addition, we list the following notation.

Mathematical Relations

- $f(x) = O(g(x))$: $|f(x)| \leq K|g(x)|$ for some constant $K > 0$ and all $x \in \mathbf{R}$.
- $f(x) = \Theta(g(x))$: $f(n) = O(g(x))$ and $g(x) = O(f(x))$.
- $f(x) = o(g(x))$ as $x \to a$: $f(x)/g(x) \to 0$ as $x \to a$.
- Adding $\tilde{}$ to all the above three means that we ignore logarithmic factors.
- $A \ll B$: $A/B \to 0$ as $n \to \infty$.
- $A \gg B$: $A/B \to \infty$ as $n \to \infty$.
- $A \sim B$: $A/B \to 1$ as some parameter converges to 0 or ∞ or another limit.
- $A \lesssim B$ or $B \gtrsim A$ if $A \leq (1 + o(1))B$.
- $[n]$: This is $\{1, 2, \ldots, n\}$. In general, if $a < b$ are positive integers, then $[a, b] = \{a, a+1, \ldots, b\}$.
- If S is a set and k is a non-negative integer, then $\binom{S}{k}$ denotes the set of k-element subsets of S. In particular, $\binom{[n]}{k}$ denotes the set of k-sets of $\{1, 2, \ldots, n\}$. Furthermore, $\binom{S}{\leq k} = \bigcup_{j=0}^{k} \binom{S}{j}$.

Graph Notation

- $G = (V, E)$: $V = V(G)$ is the vertex set and $E = E(G)$ is the edge set.
- $e(G) = |E(G)|$, and for $S \subseteq V$, we have $E_G(S) = \{e \in E : e \subseteq S\}$ and $e_G(S) = |E_G(S)|$. We drop the subscript G if the graph in question is clear.
- For $S, T \subseteq V$, let $E(S, T) = E_G(S, T) = \{\{v, w\} \in E : v \in S, w \in T\}$ and $e_G(S, T) = e(S, T) = |E_G(S, T)|$.
- $N(S) = N_G(S) = \{w \notin S : \exists v \in S$ such that $\{v, w\} \in E\}$ and $d_G(S) = |N_G(S)|$ for $S \subseteq V(G)$.
- $N_G(S, X) = N_G(S) \cap X$ for $X, S \subseteq V$.
- $d_S(x) = |\{y \in S : \{x, y\} \in E\}|$ for $x \in V, S \subseteq V$.

- For sets $X, Y \subseteq V(G)$, let $N_G(X, Y) = \{y \in Y : \exists x \in X, \{x, y\} \in E(G)\}$ and $e_G(X, Y) = |N_G(X, Y)|$.
- For $K \subseteq V(G)$ and $v \in V(G)$, let $d_K(v)$ denote the number of neighbors of v in K. The graph G is hopefully clear in the context in which this is used.
- For a graph H, $\text{aut}(H)$ denotes the number of automorphisms of H.
- $G \cong H$: graphs G and H are isomorphic.

Random Graph Models

- $[n]$: The set $\{1, 2, \ldots, n\}$.
- $\mathcal{G}_{n,m}$: The family of all labeled graphs with a vertex set $V = [n] = \{1, 2, \ldots, n\}$ and exactly m edges.
- $\mathbb{G}_{n,m}$: A random graph chosen uniformly at random from $\mathcal{G}_{n,m}$.
- $E_{n,m} = E(\mathbb{G}_{n,m})$.
- $\mathbb{G}_{n,p}$: A random graph on a vertex set $[n]$ where each possible edge occurs independently with probability p.
- $E_{n,p} = E(\mathbb{G}_{n,p})$.
- $\mathbb{G}_{n,m}^{\delta \geq k}$: $G_{n,m}$, conditioned on having a minimum degree at least k.
- $\mathbb{G}_{n,n,p}$: A random bipartite graph with a vertex set consisting of two disjoint copies of $[n]$, where each of the n^2 possible edges occurs independently with probability p.
- $\mathbb{G}_{n,r}$: A random r-regular graph on a vertex set $[n]$.
- $\mathcal{G}_{n,\mathbf{d}}$: The set of graphs with a vertex set $[n]$ and a degree sequence $\mathbf{d} = (d_1, d_2, \ldots, d_n)$.
- $\mathbb{G}_{n,\mathbf{d}}$: A random graph chosen uniformly at random from $\mathcal{G}_{n,\mathbf{d}}$.

Probability

- $\mathbb{P}(A)$: The probability of event A.
- $\mathbb{E}\, Z$: The expected value of a random variable Z.
- $h(Z)$: The entropy of a random variable Z.
- $\text{Po}(\lambda)$: A random variable with the Poisson distribution with mean λ.
- $N(0, 1)$: A random variable with the normal distribution, mean 0 and variance 1.
- $\text{Bin}(n, p)$: A random variable with the binomial distribution with the parameters n denoting the number of trials and p denoting the probability of success.
- $\text{EXP}(\lambda)$: A random variable with the exponential distribution, mean λ i.e., $\mathbb{P}(\text{EXP}(\lambda) \geq x) = e^{-\lambda x}$. We sometimes say *rate* $1/\lambda$ in place of mean λ.
- w.h.p.: A sequence of events $\mathcal{A}_n, n = 1, 2, \ldots$ is said to occur *with high probability* (w.h.p.) if $\lim_{n \to \infty} \mathbb{P}(\mathcal{A}_n) = 1$.

- $\xrightarrow{d}$: We write $X_n \xrightarrow{d} X$ to say that a random variable X_n *converges in distribution* to a random variable X, as $n \to \infty$. Occasionally we write $X_n \xrightarrow{d} N(0, 1)$ (resp. $X_n \xrightarrow{d} \text{Po}(\lambda)$) to mean that X has the corresponding normal (resp. Poisson) distribution.

Part I

Preliminaries

1 Introduction

1.1 Course Topics

In the past 30 years, random graphs, and more generally, random discrete structures, have become the focus of research of large groups of mathematicians, computer scientists, physicists, and social scientists. All these groups contribute to this area differently: mathematicians try to develop models and study their properties in a formal way while physicists and computer scientists apply those models to study real-life networks and systems and, through simulations, to develop interesting and fruitful intuitions as to how to bring mathematical models closer to reality. The abrupt development of study in the theory and applications of random graphs and networks is in a large part due to the Internet and WWW revolution, which exploded at the end of the twentieth century and, in consequence, the worldwide popularity of different social media such as Facebook or Twitter, just to name the most influential ones.

Our textbook aims to give a gentle introduction to the mathematical foundations of random graphs and to build a platform to understand the nature of real-life networks.

Although the application of probabilistic methods to prove deterministic results in combinatorics, number theory, and in other areas of mathematics have a quite long history, dating back to results of Szele and Erdős in the 1940s, the crucial step was taken by Erdős and Rényi in their seminal paper titled "On the evolution of random graphs" in 1960 (see [43]). They studied the basic properties of a large uniformly chosen random graph and studied how it evolves through the process of adding random edges, one by one. At roughly the same time, we have the important contribution of Gilbert (see [54]) in which he studied binomial random graphs where edges are inserted independently with a fixed probability. The interest in random graphs grew significantly in the mid 1980s ignited by the publication of the book by Bollobás ([21]) and due to the tireless efforts of Paul Erdős, one of the titans of twentieth-century mathematics, who was promoting probabilistic combinatorics, cooperating with mathematicians all over the world and is recognized as a founding father of the whole area. Random graphs, at the beginning of the twenty-first century is recognized as a young but quickly maturing area of mathematics, with strong connections to computer science and physics. Computer science exploits various ways of applying probabilistic concepts in the analysis of algorithms and in the construction of randomized algorithms. A common ground of random graphs and physics is particularly visible in the analysis of phase transition phenomena and in percolation. In the past 20 years, one can observe a veritable tsunami of publications dealing with various models of random graphs introduced to analyze very large real-world networks: WWW linkage, social, neural, communication, information, and transportation networks, as well as a wide range of large-scale systems.

Nowadays, research in random graphs and networks is thriving, and the subject is included in the curriculum of many mathematics and computer science departments across the world. Our textbook should help readers not only gain mathematical knowledge about the basic results of the theory of random graphs but also allow them to better understand how to model and explore real-world networks.

1.2 Course Outline

The text is divided into three parts and presents the basic elements of the theory of random graphs and networks.

To help the reader navigate through the text and to be comfortable understanding proofs, we have decided to start with describing in the preliminary part (see Chapter 2) three of the main technical tools used throughout the text. Since, in general, we look at the typical properties of large, in terms of the number n of vertices (nodes) of random graphs, in the first section of Chapter 2, we show how to deal with often complicated expressions of their numerical characteristics (random variables), in terms of their rate of growth or decline as $n \to \infty$. We next turn our attention to bounds and asymptotic approximations for factorials and binomials, frequent ingredients of the mathematical expressions found in the book. Finally, we finish this introductory, purely technical, chapter with basic information about the probabilistic tools needed to study tail bounds, i.e., probabilities that a random variable exceeds (or is smaller than) some real value. In this context, we introduce and discuss the Markov, Chebyshev, and Chernoff–Hoeffding inequalities leaving the introduction of other, more advanced, probabilistic tools to the following chapters, where they are applied for the first time.

Part II of the text is devoted to the classic Erdős–Rényi–Gilbert uniform and binomial random graphs. In Chapter 3, we formally introduce these models and discuss their relationships. We also define and study the basic features of the asymptotic behavior of random graphs, i.e., the existence of thresholds for monotone properties.

In Chapter 4, we turn our attention to the process known as the evolution of a random graph, exploring how its typical component structure evolves as the number of the edges increases one by one. We describe this process in three phases: the subcritical phase where a random graph is sparse and is a collection of small tree components and components with exactly one cycle; the phase transition, where the giant component, of order comparable with the order of random graphs, emerges; the super-critical phase, where the giant component "absorbs" smaller ones, and a random graph becomes closer and closer to the moment when it gets fully connected.

Vertex degrees, one of the most important features of random graphs, are studied in Chapter 5 in two cases: when a random graph is sparse and when it is dense. We study not only the expected values of the number of vertices of a given degree but also their asymptotic distributions, as well as applications to the notoriously difficult problem of graph isomorphism.

Chapter 6 studies the connectivity and k-connectivity of a random graph, while Chapter 7 discusses the existence in a random graph of a fixed small subgraph, whose size (the number of vertices) does not depend on the size of the random graph itself, and studies the asymptotic distribution of the number of such subgraphs in a random graph.

Large subgraphs are considered in Chapter 8. Here, the thresholds for the existence of a perfect matching are established, first for a bipartite random graph, and next, for a general random graph. These results are proved using the well-known graph theory theorems of Hall and a weakening of the corresponding theorem of Tutte, respectively. After this, long paths and cycles in sparse random graphs are studied and the proof of the celebrated result discovering the threshold for the existence of the Hamilton cycle in a random graph is given. The chapter closes with a short section on the existence of isomorphic copies of certain spanning subgraphs of random graphs.

The last chapter, in Part II, Chapter 9, is devoted to the extremes of certain graph parameters. We look first at the diameter of random graphs, i.e., the extreme value of the shortest distance between a pair of vertices. Next, we look at the size of the largest independent set and the related value of the chromatic number of a random graph.

Part III concentrates on generalizations of the Erdős–Rényi–Gilbert models of random graphs whose features better reflect some characteristic properties of real-world networks such as edge dependence, global sparseness and local clustering, small diameter, and scale-free distribution of the number of vertices of a given degree. In the first section of Chapter 10, we consider a generalization of the binomial random graph where edge probabilities, although still independent, are different for each pair of endpoints, and study conditions for its connectedness. Next, a special case of a generalized binomial random graph is introduced, where the edge probability is a function of weights of the endpoints. This is known in the literature as the Chung–Lu model. Section 12.1 provides information about the volume and uniqueness of the giant component and the sizes of other components, with respect to the expected degree sequence. The final section of Chapter 10 introduces a tool, called the configuration model, to generate a close approximation to a random graph with a fixed degree sequence. Although promoted by Bollobás, this class of random graphs is often called the Molloy–Reed model.

In Chapter 11, the "small-world" phenomenon is discussed. This name bears the observation that large real-world networks are connected by relatively short paths although being globally sparse, in the sense that the number of edges is a bounded multiple of the number of vertices, their nodes/vertices. There are two random graph models presented in this chapter: the first due to Watts and Strogatz and the second due to Kleinberg illustrate this property. In particular, finding short paths in the Kleinberg model is amenable to a particularly simple algorithm.

In general, real-world networks have a dynamic character in terms of the continual addition/deletion of vertices/edges and so we are inclined to model them via random graph processes. This is the topic of Chapter 12. There we study the properties of a wide class of preferential attachment models, which share with real networks the property that their degree sequence exhibits a tail that decays polynomially (power law), as opposed to classical random graphs, whose tails decay exponentially. We give a detailed analysis and formal description of the so-called Barabási–Albert model, as well its generalization: spatial preferential attachment.

Chapter 13 introduces the reader to the binomial and geometric random intersection graphs. Those random graphs are very useful in modeling communities with similar preferences and communication systems.

Finally, Chapter 14 is devoted to a different aspect of graph randomness. Namely, we start with a graph and equip its edges with random weights. In this chapter, we consider three of the most basic combinatorial optimization problems, namely minimum-weight spanning trees, shortest paths, and minimum weight matchings in bipartite graphs.

Suggestions for Instructors and Self-Study

The textbook material is designed for a one-semester undergraduate/graduate course for mathematics and computer science students. The course might also be recommended for students of physics, interested in networks and the evolution of large systems as well as engineering students, specializing in telecommunication. The book is almost self-contained, there being few prerequisites, although a background in elementary graph theory and probability will be helpful.

We suggest that instructors start with Chapter 2 and spend the first week with students becoming familiar with the basic rules of asymptotic computation, finding leading terms in combinatorial expressions, choosing suitable bounds for the binomials, etc., as well as probabilistic tools for tail bounds.

The core of the course is Part II, which is devoted to studying the basic properties of the classical Erdős–Rényi–Gilbert uniform and binomial random graphs. We estimate that it will take between 8 and 10 weeks to cover the material from Part II. Our suggestion for the second part of the course is to start with inhomogeneous random graphs (Chapter 10), which covers the Chung–Lu and Molloy–Reed models, continue with the "small world" (Chapter 11), and conclude with Section 12.1, i.e., the basic preferential attachment model. Any remaining time may be spent either on one of the two sections on random intersection graphs (Chapter 13), especially for those interested in social or information networks, or selected sections of Chapter 14, especially for those interested in combinatorial optimization.

To help students develop their skills in asymptotic computations, as well as to give a better understanding of the covered topics, each section of the book is concluded with simple exercises mainly of a computational nature. We ask the reader to answer rather simple questions related to the material presented in a given section. Quite often however, in particular in the sections covering more advanced topics, we just ask the reader to verify equations, developed through complicated computations, where intermediate steps have been deliberately omitted. Finally, each chapter ends with an extensive set of problems of a different scale of complication, where more challenging problems are accompanied by hints or references to the literature.

Suggestions for Further Readings

A list of possible references for books in the classical theory of random graphs is rather short. There are two advanced books in this area, the first one by Béla Bollobás [21] and the second by Svante Janson, Tomasz Łuczak, and Andrzej Ruciński [66]. Both books give a panorama of the most important problems and methods of the theory of random graphs. Since the current book is, in large part, a slimmed-down version of our earlier book [52], we encourage the reader to consult it for natural extensions of several of the topics we discuss here. Someone taking the course based on our textbook may find it helpful to refer to a very nice and friendly introduction to the theoretical aspects of random networks, in the book by Fan Chung and Linyuan Lu [32]. One may also find interesting books by Remco van der Hofstad [60] and Rick Durrett [41], which give a deep probabilistic perspective on random graphs and networks.

We may also point to an extensive literature on random networks which studies their properties via simulations, simplified heuristic analysis, experiments, and testing. Although, in general, those studies lack mathematical accuracy, they can give good intuitions and insight that help understand the nature of real-life networks. From the publications in this lively area, we would like to recommend to our reader the very extensive coverage of its problems and results presented by Mark Newman [95].

Last but not least, we suggest, in particular to someone for whom random graphs will become a favorite area of further, deeper study, reading the original paper [43] by Paul Erdős and Alfred Rényi on the evolution of random graphs, the seed from which the whole area of random graphs grew to what it is today.

2 Basic Tools

Before reading and studying results on random graphs included in the text, one should become familiar with the basic rules of asymptotic computation, find leading terms in combinatorial expressions, choose suitable bounds for the binomials, and get acquainted with probabilistic tools needed to study tail bounds, i.e., the probability that a random variable exceeds (or is smaller than) some real value. This chapter offers the reader a short description of these important technical tools used throughout the text. For more information about the topic of this chapter, we refer the reader to an excellent expository book, titled *Asymptotia*, written by Joel Spencer with Laura Florescu (see [108]).

2.1 Asymptotics

The study of random graphs and networks is mainly of an asymptotic nature. This means that we explore the behavior of discrete structures of very large "size," say n. It is quite common to analyze complicated expressions of their numerical characteristics, say $f(n)$, in terms of their rate of growth or decline as $n \to \infty$. The usual way is to "approximate" $f(n)$ with a much simpler function $g(n)$.

We say that $f(n)$ is *asymptotically equal* to $g(n)$ and write $f(n) \sim g(n)$ if $f(n)/g(n) \to 1$ as $n \to \infty$.

Example 2.1 The following functions $f(n)$ and $g(n)$ are asymptotically equal:

(a) Let $f(n) = \binom{n}{2}$, $g(n) = n^2/2$. Then $\binom{n}{2} = n(n-1)/2 \sim n^2/2$.
(b) Let $f(n) = 3\binom{n}{3}p^2$, where $p = m/\binom{n}{2}$. Find m such that $f(n) \sim g(n) = 2\omega^2$. Now,

$$f(n) = 3\frac{n(n-1)(n-2)}{6} \cdot \frac{4m^2}{(n(n-1))^2} \sim \frac{2m^2}{n},$$

so m should be chosen as $\omega\sqrt{n}$.

We write $f(n) = O(g(n))$ when there is a positive constant C such that for all sufficiently large n, $|f(n)| \leq C|g(n)|$, or, equivalently, $\limsup_{n\to\infty} |f(n)|/|g(n)| < \infty$.

Similarly, we write $f(n) = \Omega(g(n))$ when there is a positive constant c such that for all sufficiently large n, $|f(n)| \geq c|g(n)|$ or, equivalently, $\liminf_{n\to\infty} |f(n)|/|g(n)| > 0$.

Finally, we write $f(n) = \Theta(g(n))$ when there exist positive constants c and C such that for all sufficiently large n, $c|g(n)| \leq |f(n)| \leq C|g(n)|$ or, equivalently, $f(n) = O(g(n))$ and $f(n) = \Omega(g(n))$.

Note that $f(n) = O(g(n))$ simply means that the growth rate of $f(n)$ as $n \to \infty$ does not exceed the growth rate of $g(n)$, $f(n) = \Omega(g(n))$ such that $f(n)$ is growing at least as quickly as $g(n)$, while $f(n) = \Theta(g(n))$ states that their order of growth is identical.

Note also that if $f(n) = f_1(n)f_2(n) + \cdots + f_k(n)$, where k is fixed and for $i = 1, 2, \ldots, k$, $f_i(n) = O(g(n))$, then $f(n) = O(g(n))$ as well. In fact, the above property also holds if we replace O by Ω or Θ.

Example 2.2 Let

(a) $f(n) = 5n^3 - 7\log n + 2n^{-1/2}$; then

$$f(n) = O(n^3),$$
$$f(n) = 5n^3 + O(\log n),$$
$$f(n) = 5n^3 - 7\log n + O(n^{-1/2}).$$

(b) $f(x) = e^x$; then $f(x) = 1 + x + x^2/2 + O(x^3)$ for $x \to 0$.

We now introduce the frequently used "little o" notation.

We write $f(n) = o(g(n))$ or $f(n) \ll g(n)$ if $f(n)/g(n) \to 0$ as $n \to \infty$. Similarly, we write $f(n) = \omega(g(n))$ or $f(n) \gg g(n)$ if $f(n)/g(n) \to \infty$ as $n \to \infty$. Obviously, if $f(n) \ll g(n)$, then we can also write $g(n) \gg f(n)$.

Note that $f(n) = o(g(n))$ simply means that $g(n)$ grows faster with n than $f(n)$, and the other way around if $f(n) = \omega(g(n))$.

Let us also make a few important observations. Obviously, $f(n) = o(1)$ means that $f(n)$ itself tends to 0 as $n \to \infty$. Also the notation $f(n) \sim g(n)$ is equivalent to the statement that $f(n) = (1 + o(1))g(n)$. One should also note the difference between the $(1 + o(1))$ factor in the expression $f(n) = (1 + o(1))g(n)$ and when it is placed in the exponent, i.e., when $f(n) = g(n)^{1+o(1)}$. In the latter case, this notation means that for every fixed $\varepsilon > 0$ and sufficiently large n, $g(n)^{1-\varepsilon} < f(n) < g(n)^{1+\varepsilon}$. Hence here the $(1 + o(1))$ factor is more accurate in $f(n) = (1 + o(1))g(n)$ than the much coarser factor $(1 + o(1))$ in $f(n) = g(n)^{(1+o(1))}$.

It is also worth mentioning that, regardless of how small a constant $c > 0$ is and however large a positive constant C is, the following hierarchy of growths holds:

$$\ln^C n \ll n^c, \quad n^C \ll (1+c)^n, \quad C^n \ll n^{cn}. \tag{2.1}$$

Example 2.3 Let

(a) $f(n) = \binom{n}{2}p$, where $p = p(n)$. Then $f(n) = o(1)$ if $p = 1/n^{2+\varepsilon}$, where $\varepsilon > 0$, since $f(n) \sim \frac{n^2}{2}n^{-2-\varepsilon} = n^{-\varepsilon}/2 \to 0$.

(b) $f(n) = 3\binom{n}{3}p^2$, where $p = m/\binom{n}{2}$ and $m = n^{1/2}/\omega$, where $\omega = \omega(n) \to \infty$ as $n \to \infty$. Then $f(n) = o(1)$ since $f(n) \le n^4/\left(2\binom{n}{2}^2\omega^2\right) \to 0$.

Exercises

2.1.1 Let $f(n) = \binom{n}{3}p^2(1-p)^{2(n-3)}$, where $\log n - \log\log n \le np \le 2\log n$. Then show that $f(n) = O(n^3p^2e^{-2np}) = o(1)$.

2.1.2 Let $f(n) = \left(1 - \frac{c}{n}\right)^{n(\log n)^2}$, where c is a constant. Then show that $f(n) = o(n^{-2})$.

2.1.3 Let $p = \frac{\log n}{n}$ and $f(n) = \sum_{k=2}^{n/2} n^k p^{k-1}(1-p)^{k(n-k)}$. Then show that $f(n) = o(1)$.

2.1.4 Suppose that $k = k(n) = \lceil 2\log_{1/(1-p)} n\rceil$ and $0 < p < 1$ is a constant. Then show that $\binom{n}{k}(1-p)^{k(k-1)/2} \to 0$.

2.2 Binomials

We start with the famous asymptotic estimate for $n!$, known as Stirling's formula.

Lemma 2.4

$$n! = (1 + o(1))n^n e^{-n}\sqrt{2\pi n}.$$

Moreover,

$$n^n e^{-n}\sqrt{2\pi n} \le n! \le n^n e^{-n}\sqrt{2\pi n}\, e^{1/12n}.$$

Example 2.5 Consider the coin-tossing experiment where we toss a fair coin $2n$ times. What is the probability that this experiment results in *exactly* n heads and n tails? Let A denote such an event. Then $\mathbb{P}(A) = \binom{2n}{n}2^{-2n}$.

By Stirling's approximation,

$$\binom{2n}{n} = \frac{(2n)!}{(n!)^2} \sim \frac{(2n)^{2n}e^{-2n}\sqrt{2\pi(2n)}}{(n^n e^{-n})^2(2\pi n)} = \frac{2^{2n}}{\sqrt{\pi n}}.$$

Hence $\mathbb{P}(A) \sim 1/\sqrt{\pi n}$.

Example 2.6 What is the number of digits in 100!? To answer this question we shall use sharp bounds on $n!$ given in Lemma 2.4. Notice that

$$1 < \frac{n!}{n^n e^{-n}\sqrt{2\pi n}} < e^{1/12n}.$$

Hence, taking logarithms,

$$0 < \ln n! - \left(n + \frac{1}{2}\right) \ln n - n + \frac{1}{2} \ln 2\pi < \frac{1}{12n}.$$

Now, since the number of digits in a positive integer n is $\lfloor \log_{10} n + 1 \rfloor$, we divide both sides by ln 10, to get

$$0 < \log_{10} n! - \frac{\left(n + \frac{1}{2}\right) \ln n - n + \frac{1}{2} \ln 2\pi}{\ln 10} < \frac{1}{12n \ln 10}.$$

Substituting $n = 100$, we obtain

$$0 < \log_{10} 100! - 157.96 < 0.00036.$$

Thus 100! has exactly 158 digits.

Before we move to the analysis of the asymptotic behavior of the binomial coefficient $\binom{n}{k}$, let us prove some simple, often-used upper and lower bounds, valid for all fixed n and k.

Lemma 2.7 *For every integer n and k, $k \leq n$,*

$$\binom{n}{k} \leq \left(\frac{ne}{k}\right)^k, \tag{2.2}$$

$$\frac{n^k}{k!}\left(1 - \frac{k(k-1)}{2n}\right) \leq \binom{n}{k} \leq \frac{n^k}{k!}\left(1 - \frac{k}{2n}\right)^{k-1}, \tag{2.3}$$

$$\binom{n}{k} \leq \frac{n^k}{k!} e^{-k(k-1)/(2n)}. \tag{2.4}$$

Proof To prove (2.2), note that

$$\binom{n}{k} = \frac{n!}{k!(n-k)!} = \frac{(n)_k}{k!},$$

where

$$(n)_k = n(n-1)(n-2)\cdots(n-k+1) \leq n^k.$$

By Lemma 2.4, $k! > (k/e)^k$ and the first bound holds.

To see that remaining bounds on $\binom{n}{k}$ are true we have to estimate $(n)_k$ more carefully. Note first that

$$\binom{n}{k} = \frac{n^k}{k!} \frac{(n)_k}{n^k} = \frac{n^k}{k!} \prod_{i=0}^{k-1} \left(1 - \frac{i}{n}\right). \tag{2.5}$$

The upper bound in (2.3) follows from the observation that for $i = 1, 2, \ldots, \lfloor k/2 \rfloor$,

$$\left(1 - \frac{i}{n}\right)\left(1 - \frac{k-i}{n}\right) \leq \left(1 - \frac{k}{2n}\right)^2.$$

The lower bound in (2.3) is implied by the Weierstrass product inequality, which states that

$$\prod_{r=1}^{s}(1 - a_r) + \sum_{r=1}^{s} a_r \geq 1 \tag{2.6}$$

for $0 \leq a_1, a_2, \ldots, a_s \leq 1$, and can be easily proved by induction. Hence

$$\prod_{i=0}^{i-1}\left(1 - \frac{i}{n}\right) \geq 1 - \sum_{i=0}^{k-1} \frac{i}{n} = 1 - \frac{k(k-1)}{2n}.$$

The last bound given in (2.4) immediately follows from the upper bound in (2.3) and the simple observation that for every real x,

$$1 + x \leq e^x. \tag{2.7}$$

□

Example 2.8 To illustrate an application of (2.2), let us consider the function

$$f(n,k) = \binom{n}{k}(1 - 2^{-k})^n,$$

where n, k are positive integers, and denote by n_k the smallest n (as a function of k) such that $f(n,k) < 1$. We aim for an upper estimate of n_k as a function of k, when $k \geq 2$. In fact, we claim that

$$n_k \leq \left(1 + \frac{3\log_2 k}{k}\right) k^2 2^k \ln 2. \tag{2.8}$$

Now, by (2.2) and (2.7),

$$f(n,k) = \binom{n}{k}(1 - 2^{-k})^n \leq \left(\frac{ne}{k}\right)^k e^{-n/2^k}.$$

If $m = (1 + \varepsilon)k^2 2^k \ln 2$, then

$$\left(\frac{me}{k}\right)^k e^{-m/2^k} = ((1 + \varepsilon)2^k k 2^{-(1+\varepsilon)k} e \ln 2)^k. \tag{2.9}$$

If $\varepsilon = 3\log_2 k/k$, then the right-hand side (RHS) of (2.9) equals $((1 + \varepsilon)k^{-2} e \ln 2)^k$, which is less than 1. This implies that n_k satisfies (2.8).

In the following chapters, we shall also need the bounds given in the next lemma.

Lemma 2.9 *If $a \geq b$, then*

$$\left(\frac{k-b}{n-b}\right)^b \left(\frac{n-k-a+b}{n-a}\right)^{a-b} \leq \frac{\binom{n-a}{k-b}}{\binom{n}{k}} \leq \left(\frac{k}{n}\right)^b \left(\frac{n-k}{n-b}\right)^{a-b}.$$

Proof To see this note that

$$\frac{\binom{n-a}{k-b}}{\binom{n}{k}} = \frac{(n-a)!k!(n-k)!}{(k-b)!(n-k-a+b)!n!}$$
$$= \frac{k(k-1)\cdots(k-b+1)}{n(n-1)\cdots(n-b+1)} \times \frac{(n-k)(n-k-1)\cdots(n-k-a+b+1)}{(n-b)(n-b-1)\cdots(n-a+1)}$$
$$\leq \left(\frac{k}{n}\right)^b \left(\frac{n-k}{n-b}\right)^{a-b}.$$

The lower bound follows similarly. □

Example 2.10 Let us show that

$$\frac{\binom{\binom{n}{2}-2l+r}{m-2l+r}}{\binom{\binom{n}{2}}{m}} = O\left(\frac{(2m)^{2l-r}}{n^{4l-2r}}\right)$$

assuming that $2l - r \ll m, n$.

Applying Lemma 2.9 with n replaced by $\binom{n}{2}$ and with $k = m, a = b = 2l - r$, we see that

$$\frac{\binom{\binom{n}{2}-2l+r}{m-2l+r}}{\binom{\binom{n}{2}}{m}} \leq \left(\frac{m}{\binom{n}{2}}\right)^{2l-r}.$$

We will also need precise estimates for the binomial coefficient $\binom{n}{k}$ when $k = k(n)$. They are based on the Stirling approximation of factorials and estimates given in Lemma 2.7.

Lemma 2.11 *Let k be fixed or grow with n as $n \to \infty$. Then*

$$\binom{n}{k} \sim \frac{n^k}{k!} \quad \text{if } k = o(n^{1/2}), \tag{2.10}$$

$$\binom{n}{k} \sim \frac{n^k}{k!} \exp\left\{-\frac{k^2}{2n}\right\} \quad \text{if } k = o(n^{2/3}), \tag{2.11}$$

$$\binom{n}{k} \sim \frac{n^k}{k!} \exp\left\{-\frac{k^2}{2n} - \frac{k^3}{6n^2}\right\} \quad \text{if } k = o(n^{3/4}). \tag{2.12}$$

Proof The asymptotic formula (2.10) follows directly from (2.3) and (2.4). We only prove (2.12) since the proof of (2.11) is analogous. In fact, in the proofs of these bounds we use the Taylor expansion of $\ln(1-x),\ 0 < x < 1$. In the case of (2.12), we take

$$\ln(1-x) = -x - \frac{x^2}{2} + O(x^3). \tag{2.13}$$

Now,

$$\begin{aligned}\binom{n}{k} &= \frac{n^k}{k!}\prod_{i=0}^{k-1}\left(1-\frac{i}{n}\right)\\ &= \frac{n^k}{k!}\exp\left\{\ln\left[\prod_{i=0}^{k-1}\left(1-\frac{i}{n}\right)\right]\right\}\\ &= \frac{n^k}{k!}\exp\left\{\sum_{i=0}^{k-1}\ln\left(1-\frac{i}{n}\right)\right\}\\ &= \frac{n^k}{k!}\exp\left\{-\sum_{i=0}^{k-1}\left(\frac{i}{n}+\frac{i^2}{2n^2}\right)+O\left(\frac{k^4}{n^3}\right)\right\}.\end{aligned}$$

Hence

$$\binom{n}{k} = \frac{n^k}{k!}\exp\left\{-\frac{k^2}{2n}-\frac{k^3}{6n^2}+O\left(\frac{k^4}{n^3}\right)\right\}, \tag{2.14}$$

and equation (2.12) follows. □

Example 2.12 Let n be a positive integer, $k = o(n^{1/2})$ and $m = o(n)$. Applying (2.10) and the bounds from Lemma 2.9 we show that

$$\frac{1}{2}\binom{n}{k}(k-1)!\frac{\binom{\binom{n}{2}-k}{m-k}}{\binom{\binom{n}{2}}{m}} \sim \frac{1}{2}\frac{n^k}{k!}(k-1)!\left(\frac{2m}{n^2}\right)^k \sim \frac{\left(\frac{2m}{n}\right)^k}{2k}.$$

Example 2.13 As an illustration of the application of (2.11) we show that if $k = k(n) \gg n^{2/5}$, then

$$f(n,k) = \binom{n}{k}k^{k-2}\left(\frac{1}{n}\right)^{k-1}\left(1-\frac{1}{n}\right)^{\binom{k}{2}-k+1+k(n-k)} = o(1).$$

By (2.11) and Stirling's approximation of $k!$ (Lemma 2.4), we get

$$\binom{n}{k} \sim \frac{n^k}{k!}e^{-k^2/2n} \sim e^{-k^2/2n}\left(\frac{ne}{k}\right)^k(2\pi k)^{-1/2}.$$

Moreover, since

$$\binom{k}{2}-k+1+k(n-k) = kn-\frac{k^2}{2}+O(k)$$

and

$$\ln\left(1-\frac{1}{n}\right) = -\frac{1}{n}+O(n^{-2}),$$

we have

$$\left(1-\frac{1}{n}\right)^{\binom{k}{2}-k+1+k(n-k)} = \exp\left\{-k+\frac{k^2}{2n}+o(1)\right\}.$$

Hence

$$f(n,k) \sim e^{-k^2/2n}\left(\frac{ne}{k}\right)^k k^{k-2}(2\pi k)^{-1/2} n^{-k+1} e^{-k+k^2/2n}$$
$$\sim nk^{-5/2}(2\pi)^{-1/2} = o(1).$$

Example 2.14 Let

$$f(n,k,l) = \binom{n}{k} C(k,k+l) p^{k+l}(1-p)^{\binom{k}{2}-(k+l)+k(n-k)}, \tag{2.15}$$

where $k \le n,\ l = o(k)$ and $np = 1+\varepsilon,\ 0 < \varepsilon < 1$.

Assuming that $f(n,k,l) \le n/k$ and applying (2.14) and (2.13), we look for an asymptotic upper bound on $C(k,k+l)$ as follows:

$$f(n,k,l) = C(k,k+l)p^{k+l}\frac{n^k}{k!}\exp\left\{-\frac{k^2}{2n} - \frac{k^3}{6n^2} + O\left(\frac{k^4}{n^3}\right)\right\}$$
$$\times \exp\left\{\left(-p - \frac{p^2}{2} + O(p^3)\right)\left(\binom{k}{2} - (k+l) + k(n-k)\right)\right\}$$
$$= C(k,k+l)\frac{(np)^{k+l}}{n^l k!}\exp\left\{-\frac{k^2}{2n} - \frac{k^3}{6n^2} - pkn + \frac{pk^2}{2}\right\}$$
$$\times \exp\left\{O\left(\frac{k^4}{n^3} + pk + p^2 kn\right)\right\}.$$

Recalling that $f(n,k,l) \le n/k,\ p = (1+\varepsilon)/n$ and using the Stirling approximation for $k!$, we get

$$C(k,k+l) \le n^{l+1}(k-1)!\exp\left\{-\varepsilon k + \frac{\varepsilon^2 k}{2} + \frac{k^3}{6n^2} + k + \varepsilon k - \frac{\varepsilon k^2}{2n}\right\}$$
$$\times \exp\left\{O\left(\frac{k^4}{n^3} + \varepsilon l\right)\right\}$$
$$\le 3n^{l+1}k^{k-\frac{1}{2}}\exp\left\{\frac{\varepsilon^2 k}{2} + \frac{k^3}{6n^2} - \frac{\varepsilon k^2}{2n} + O\left(\frac{k^4}{n^3} + \varepsilon l\right)\right\}.$$

Exercises

2.2.1 Let $f(n,k) = 2\frac{\binom{k}{2}\binom{n-k}{k-2}}{\binom{n}{k}}$, where $k = k(n) \to \infty$ as $n \to \infty,\ k = o(n^{1/2})$. Show that $f(n,k) \sim k^4/n^2$.

2.2.2 Let $f(n,k) = \left(\frac{\binom{n-2}{k}}{\binom{n-1}{k}}\right)^{n-1}$, where $k = k(n)$. Show that $f(n,k) \sim e^{-k}$.

2.2.3 Let $f(n,k) = \sum_{k=2}^{n}\sum_{j=0}^{n-k} k^2\binom{n}{k}\binom{n}{j}(k-1)!j!\left(\frac{c}{n}\right)^{k+j+1}$, where $c < 1$. Show that $f(n,k) = O(1/n)$.

2.2.4 Apply (2.2) to show that

$$f(n,k) = \sum_{k=1}^{n/1000} \binom{n}{k}\binom{n}{2k}\left(\frac{\binom{3k}{2}}{\binom{n-1}{2}}\right)^{2k} = o(1).$$

2.2.5 Prove that n_k in Example 2.8 satisfies $n_k \geq k^2 2^k \ln k$ for sufficiently large k. Use equation (2.3) and $\ln(1-x) \geq -\frac{x}{1-x}$ if $0 < x < 1$ to get a lower bound for $(1-2^{-k})^n$. The latter inequality following from $\ln(1-x) = -\sum_{n=1}^{\infty} \frac{x^n}{n}$.

2.3 Tail Bounds

One of the most basic and useful tools in the study of random graphs is *tail bounds*, i.e., upper bounds on the probability that a random variable exceeds a certain real value. We first explore the potential of the simple but indeed very powerful Markov inequality.

Lemma 2.15 (Markov Inequality) *Let X be a non-negative random variable. Then, for all $t > 0$,*

$$\mathbb{P}(X \geq t) \leq \frac{\mathbb{E}X}{t}.$$

Proof Let

$$I_A = \begin{cases} 1 & \text{if event } A \text{ occurs,} \\ 0 & \text{otherwise.} \end{cases}$$

Notice that

$$X = XI_{\{X \geq t\}} + XI_{\{X<t\}} \geq XI_{\{X \geq t\}} \geq tI_{\{X \geq t\}}.$$

Hence,

$$\mathbb{E}X \geq t\mathbb{E}I_{\{X \geq t\}} = t\,\mathbb{P}(X \geq t).$$

□

Example 2.16 Let X be a random variable with the expectation $\mathbb{E}\,X = n((n-2)/n))^m$, where $m = m(n)$. Find m such that

$$\mathbb{P}(X \geq \sqrt{n}) \leq e^{-c},$$

where $c > 0$ is a constant. By the Markov inequality

$$\mathbb{P}(X \geq \sqrt{n}) \leq \frac{n\left(\frac{n-2}{n}\right)^m}{\sqrt{n}} \leq \sqrt{n}e^{-2m/n}.$$

So m should be chosen as $m = \frac{1}{2}n(\log n^{1/2} + c)$.

Example 2.17 Let X be a random variable with the expectation

$$\mathbb{E}\,X_k = \binom{n}{k} k^{k-2} p^{k-1},$$

where $k \geq 3$ is fixed. Find $p = p(n)$ such that $\mathbb{P}(X \geq 1) = O(\omega^{1-k})$, where $\omega = \omega(n)$. Note that by the Markov inequality $\mathbb{P}(X \geq 1) \leq \mathbb{E}\,X$; hence

$$\mathbb{P}(X \geq 1) \leq \binom{n}{k} k^{k-2} p^{k-1} \leq \left(\frac{ne}{k}\right)^k k^{k-2} p^{k-1}.$$

Now put $p = 1/(\omega n^{k/(k-1)})$ to get

$$\mathbb{P}(X \geq 1) \leq \left(\frac{ne}{k}\right)^k k^{k-2} \left(\frac{1}{\omega n^{k/(k-1)}}\right)^{k-1} = \frac{e^k}{k^2 \omega^{k-1}} = O(\omega^{1-k}).$$

We are very often concerned with bounds on the *upper and lower tail* of the distribution of S, i.e., on $\mathbb{P}(X \geq \mathbb{E}\,X + t)$ and $\mathbb{P}(X \leq \mathbb{E}\,X - t)$, respectively. The following joint tail bound on the deviation of a random variable from its expectation is a simple consequence of Lemma 2.15.

Lemma 2.18 (Chebyshev Inequality) *If X is a random variable with a finite mean and variance, then, for $t > 0$,*

$$\mathbb{P}(|X - \mathbb{E}\,X| \geq t) \leq \frac{\operatorname{Var} X}{t^2}.$$

Proof

$$\mathbb{P}(|X - \mathbb{E}\,X| \geq t) = \mathbb{P}((X - \mathbb{E}\,X)^2 \geq t^2) \leq \frac{\mathbb{E}(X - \mathbb{E}\,X)^2}{t^2} = \frac{\operatorname{Var} X}{t^2}.$$

□

Example 2.19 Consider a standard coin-tossing experiment where we toss a fair coin n times and count, say, the number X of heads. Note that $\mu = \mathbb{E}\,X = n/2$, while $\operatorname{Var} X = n/4$. So, by the Chebyshev inequality,

$$\mathbb{P}\left(\left|X - \frac{n}{2}\right| \geq \varepsilon n\right) \leq \frac{n/4}{(\varepsilon n)^2} = \frac{1}{4n\varepsilon^2}.$$

Hence,

$$\mathbb{P}\left(\left|\frac{X}{n} - \frac{1}{2}\right| \geq \varepsilon\right) \leq \frac{1}{4n\varepsilon^2},$$

so if we choose, for example, $\varepsilon = 1/4$, we get the following bound:

$$\mathbb{P}\left(\left|\frac{X}{n} - \frac{1}{2}\right| \geq \frac{1}{4}\right) \leq \frac{4}{n}.$$

Suppose again that X is a random variable and $t > 0$ is a real number. We focus our attention on the observation due to Bernstein [17], which can lead to the derivation of stronger bounds on the lower and upper tails of the distribution of the random variable X.

Let $\lambda \geq 0$ and $\mu = \mathbb{E}\, X$; then

$$\mathbb{P}(X \geq \mu + t) = \mathbb{P}(e^{\lambda X} \geq e^{\lambda(\mu+t)}) \leq e^{-\lambda(\mu+t)}\, \mathbb{E}(e^{\lambda X}) \tag{2.16}$$

by the Markov inequality (see Lemma 2.15).

Similarly for $\lambda \leq 0$,

$$\mathbb{P}(X \leq \mu - t) = \mathbb{P}(e^{\lambda X} \geq e^{\lambda(\mu-t)}) \leq e^{-\lambda(\mu-t)}\, \mathbb{E}(e^{\lambda X}). \tag{2.17}$$

Combining (2.16) and (2.17) one can obtain a bound for $\mathbb{P}(|X - \mu| \geq t)$. A bound of such type was considered above, that is, the Chebyshev inequality.

We will next discuss in detail tail bounds for the case where a random variable is the sum of independent random variables. This is a common case in the theory of random graphs. Let

$$S_n = X_1 + X_2 + \cdots + X_n,$$

where $X_i, i = 1, \ldots, n$ are independent random variables.

Assume that $0 \leq X_i \leq 1$ and $\mathbb{E}\, X_i = \mu_i$ for $i = 1, 2, \ldots, n$. Let $\mathbb{E}\, S_n = \mu_1 + \mu_2 + \cdots + \mu_n = \mu$. Then, by (2.16), for $\lambda \geq 0$,

$$\mathbb{P}(S_n \geq \mu + t) \leq e^{-\lambda(\mu+t)} \prod_{i=1}^{n} \mathbb{E}(e^{\lambda X_i}), \tag{2.18}$$

and, by (2.16), for $\lambda \leq 0$,

$$\mathbb{P}(S_n \leq \mu - t) \leq e^{-\lambda(\mu-t)} \prod_{i=1}^{n} \mathbb{E}(e^{\lambda X_i}). \tag{2.19}$$

In the above bounds we applied the observation that the expected value of the product of independent random variables is equal to the product of their expectations. Note also that $\mathbb{E}(e^{\lambda X_i})$ in (2.18) and (2.19), likewise $\mathbb{E}(e^{\lambda X})$ in (2.16) and (2.17), are the moment-generating functions of the X_i and X, respectively. So finding bounds boils down to the estimation of these functions. Now the convexity of e^x and $0 \leq X_i \leq 1$ implies that

$$e^{\lambda X_i} \leq 1 - X_i + X_i e^{\lambda}.$$

Taking expectations, we get

$$\mathbb{E}(e^{\lambda X_i}) \leq 1 - \mu_i + \mu_i e^{\lambda}.$$

Equation (2.18) becomes, for $\lambda \geq 0$,

$$\begin{aligned} \mathbb{P}(S_n \geq \mu + t) &\leq e^{-\lambda(\mu+t)} \prod_{i=1}^{n} (1 - \mu_i + \mu_i e^{\lambda}) \\ &\leq e^{-\lambda(\mu+t)} \left(\frac{n - \mu + \mu e^{\lambda}}{n} \right)^n. \end{aligned} \tag{2.20}$$

The second inequality follows from the fact that the geometric mean is at most the arithmetic mean, i.e., $(x_1 x_2 \cdots x_n)^{1/n} \leq (x_1+x_2+\cdots+x_n)/n$ for non-negative $x_1, x_2, \dots, x_n$. This in turn follows from Jensen's inequality and the concavity of $\log x$. The RHS of (2.20) attains its minimum, as a function of λ, at

$$e^{\lambda} = \frac{(\mu+t)(n-\mu)}{(n-\mu-t)\mu}. \tag{2.21}$$

Hence, by (2.20) and (2.21), assuming that $\mu + t < n$,

$$\mathbb{P}(S_n \geq \mu + t) \leq \left(\frac{\mu}{\mu+t}\right)^{\mu+t} \left(\frac{n-\mu}{n-\mu-t}\right)^{n-\mu-t}, \tag{2.22}$$

while for $t > n - \mu$ this probability is zero.

Now let

$$\begin{aligned} \varphi(x) &= (1+x)\log(1+x) - x, \quad x \geq -1, \\ &= \sum_{k=2}^{\infty} \frac{(-1)^k x^k}{k(k-1)} \quad \text{for } |x| \leq 1, \end{aligned}$$

and let $\varphi(x) = \infty$ for $x < -1$. Now, for $0 \leq t < n - \mu$, we can rewrite the bound (2.22) as

$$\mathbb{P}(S_n \geq \mu + t) \leq \exp\left\{-\mu\varphi\left(\frac{t}{\mu}\right) - (n-\mu)\varphi\left(\frac{-t}{n-\mu}\right)\right\}.$$

Since $\varphi(x) \geq 0$ for every x, we get

$$\mathbb{P}(S_n \geq \mu + t) \leq e^{-\mu\varphi(t/\mu)}. \tag{2.23}$$

Similarly, putting $n - S_n$ for S_n, or by an analogous argument, using (2.19), we get, for $0 \leq t \leq \mu$,

$$\mathbb{P}(S_n \leq \mu - t) \leq \exp\left\{-\mu\varphi\left(\frac{-t}{\mu}\right) - (n-\mu)\varphi\left(\frac{t}{n-\mu}\right)\right\}.$$

Hence,

$$\mathbb{P}(S_n \leq \mu - t) \leq e^{-\mu\varphi(-t/\mu)}. \tag{2.24}$$

We can simplify expressions (2.23) and (2.24) by observing that

$$\varphi(x) \geq \frac{x^2}{2(1+x/3)}. \tag{2.25}$$

To see this observe that for $|x| \leq 1$, we have

$$\varphi(x) - \frac{x^2}{2(1+x/3)} = \sum_{k=2}^{\infty} (-1)^k \left(\frac{1}{k(k-1)} - \frac{1}{2 \cdot 3^{k-2}}\right) x^k.$$

Equation (2.25) for $|x| \leq 1$ follows from $\frac{1}{k(k-1)} - \frac{1}{2 \cdot 3^{k-2}} \geq 0$ for $k \geq 2$. We leave it as Exercise 2.3.3 to check that (2.25) remains true for $x > 1$.

Taking this into account we arrive at the following theorem, see Hoeffding [59].

Theorem 2.20 (Chernoff–Hoeffding inequality) *Suppose that* $S_n = X_1 + X_2 + \cdots + X_n$ *while, for* $i = 1, 2, \ldots, n$,

(i) $0 \le X_i \le 1$,
(ii) $X_1, X_2, \ldots, X_n$ *are independent.*

Let $\mathbb{E}\, X_i = \mu_i$ *and* $\mu = \mu_1 + \mu_2 + \cdots + \mu_n$. *Then for* $t \ge 0$,

$$\mathbb{P}(S_n \ge \mu + t) \le \exp\left\{-\frac{t^2}{2(\mu + t/3)}\right\} \tag{2.26}$$

and for $t \le \mu$,

$$\mathbb{P}(S_n \le \mu - t) \le \exp\left\{-\frac{t^2}{2(\mu - t/3)}\right\}. \tag{2.27}$$

Putting $t = \varepsilon\mu$, for $0 < \varepsilon < 1$, in (2.23), (2.26) and (2.27), one can immediately obtain the following bounds.

Corollary 2.21 *Let* $0 < \varepsilon < 1$; *then*

$$\mathbb{P}(S_n \ge (1+\varepsilon)\mu) \le \left(\frac{e^{\varepsilon}}{(1+\varepsilon)^{1+\varepsilon}}\right)^{\mu} \le \exp\left\{-\frac{\mu\varepsilon^2}{3}\right\}, \tag{2.28}$$

while

$$\mathbb{P}(S_n \le (1-\varepsilon)\mu) \le \exp\left\{-\frac{\mu\varepsilon^2}{2}\right\}. \tag{2.29}$$

□

Note also that the bounds (2.28) and (2.29) imply that, for $0 < \varepsilon < 1$,

$$\mathbb{P}(|S_n - \mu| \ge \varepsilon\mu) \le 2\exp\left\{-\frac{\mu\varepsilon^2}{3}\right\}. \tag{2.30}$$

Example 2.22 Let us return to the coin-tossing experiment from Example 2.19. Notice that the number of heads X is in fact the sum of binary random variables X_i, for $i = 1, 2, \ldots, n$, each representing the result of a single experiment, that is, $X_i = 1$, with probability $1/2$, when head occurs in the ith experiment, and $X_i = 0$, with probability $1/2$, otherwise. Denote this sum by $S_n = X_1 + X_2 + \cdots + X_n$ and notice that random variables X_i are independent. Applying the Chernoff bound (2.30), we get

$$\mathbb{P}\left(\left|S_n - \frac{n}{2}\right| \ge \varepsilon\frac{n}{2}\right) \le 2\exp\left\{-\frac{n\varepsilon^2}{6}\right\}.$$

Choosing $\varepsilon = 1/2$, we get

$$\mathbb{P}\left(\left|\frac{S_n}{n} - \frac{1}{2}\right| \geq \frac{1}{4}\right) \leq 2e^{-n/24},$$

a huge improvement over the Chebyshev bound.

Example 2.23 Let S_n now denote the number of heads minus the number of tails after n flips of a fair coin. Find $\mathbb{P}(S_n \geq \omega\sqrt{n})$, where $\omega = \omega(n) \to \infty$ arbitrarily slowly, as $n \to \infty$.

Notice that S_n is again the sum of independent random variables X_i, but now $X_i = 1$, with probability $1/2$, when head occurs in the ith experiment, while $X_i = -1$, with probability $1/2$, when tail occurs. Hence, for each $i = 1, 2, \ldots, n$, expectation $\mathbb{E}\, X_i = 0$ and variance $\operatorname{Var} X_i = 1$. Therefore, $\mu = \mathbb{E}\, S_n = 0$ and $\sigma^2 = \operatorname{Var} S_n = n$. So, by (2.26)

$$\mathbb{P}(S_n \geq \omega\sqrt{n}) \leq e^{-3\omega/2}.$$

To compare, notice that Chebyshev's inequality yields the much weaker bound since it implies that

$$\mathbb{P}(S_n \geq \omega\sqrt{n}) \leq \frac{1}{2\omega^2}.$$

One can "tailor" the Chernoff bounds with respect to specific needs. For example, for small ratios t/μ, the exponent in (2.26) is close to $t^2/2\mu$, and the following bound holds.

Corollary 2.24

$$\mathbb{P}(S_n \geq \mu + t) \leq \exp\left\{-\frac{t^2}{2\mu} + \frac{t^3}{6\mu^2}\right\} \tag{2.31}$$

$$\leq \exp\left\{-\frac{t^2}{3\mu}\right\} \qquad \textit{for } t \leq \mu. \tag{2.32}$$

Proof Use (2.26) and note that

$$(\mu + t/3)^{-1} \geq (\mu - t/3)/\mu^2.$$

□

Example 2.25 Suppose that $p = c/n$ for some constant c and that we create an $n \times n$ matrix A with values 0 or 1, where for all i, j, $\Pr(A(i, j) = 1) = p$ independently of other matrix entries. Let Z denote the number of columns that are all zero. We will show that, for small $\varepsilon > 0$,

$$\Pr(Z \geq (1 + \varepsilon)e^{-c}n) \leq e^{-\varepsilon^2 e^{-c} n/3}.$$

Each column of A is zero with probability $q = (1-p)^n = (1+O(1/n))e^{-c}$. Furthermore, Z is the sum of indicator random variables and is distributed as the binomial $Bin(n,q)$. Applying (2.31) with $\mu = nq, t = \varepsilon\mu$, we get

$$\Pr(Z \geq (1+\varepsilon)e^{-c}n) \leq \exp\left\{-\frac{\varepsilon^2\mu}{2} + \frac{\varepsilon^3\mu}{6}\right\} \leq \exp\left\{-\frac{\varepsilon^2 e^{-c} n}{3}\right\}.$$

For large deviations we have the following result.

Corollary 2.26 *If $c > 1$, then*

$$\mathbb{P}(S_n \geq c\mu) \leq \left(\frac{e}{ce^{1/c}}\right)^{c\mu}. \tag{2.33}$$

Proof Put $t = (c-1)\mu$ into (2.23). □

Example 2.27 Let $X_1, X_2, \ldots, X_n$ be independent binary random variables, that is, $X_i \in \{0,1\}$ with the Bernoulli distribution: $\mathbb{P}(X_i = 1) = p$, $\mathbb{P}(X_i = 0) = 1-p$, for every $1 \leq i \leq n$, where $0 < p < 1$. Then $S_n = \sum_{i=1}^n X_i$ has the binomial distribution with the expectation $\mathbb{E}\, S_n = \mu = np$. Applying Corollary 2.26 one can easily show that for $t = 2e\mu$,

$$\mathbb{P}(S_n \geq t) \leq 2^{-t}.$$

Indeed, for $c = 2e$,

$$\mathbb{P}(S_n \geq t) = \mathbb{P}(S_n \geq c\mu) \leq \left(\frac{e}{ce^{1/c}}\right)^{c\mu} \leq \left(\frac{1}{2e^{1/(2e)}}\right)^{2e\mu} \leq 2^{-t}.$$

We also have the following:

Corollary 2.28 *Suppose that $X_1, X_2, \ldots, X_n$ are independent random variables and that $a_i \leq X_i \leq b_i$ for $i = 1, 2, \ldots, n$. Let $S_n = X_1 + X_2 + \cdots + X_n$ and $\mu_i = \mathbb{E}(X_i)$, $i = 1, 2, \ldots, n$ and $\mu = \mathbb{E}(S_n)$. Then for $t > 0$ and $c_i = b_i - a_i$, $i = 1, 2, \ldots, n$, we have*

$$\mathbb{P}(S_n \geq \mu + t) \leq \exp\left\{-\frac{2t^2}{c_1^2 + c_2^2 + \cdots + c_n^2}\right\}, \tag{2.34}$$

$$\mathbb{P}(S_n \leq \mu - t) \leq \exp\left\{-\frac{2t^2}{c_1^2 + c_2^2 + \cdots + c_n^2}\right\}. \tag{2.35}$$

Proof We can assume without loss of generality that $a_i = 0, i = 1, 2, \ldots, n$. We just subtract $A = \sum_{i=1}^n a_i$ from S_n. We proceed as before.

$$\mathbb{P}(S_n \geq \mu + t) = \mathbb{P}\left(e^{\lambda S_n} \geq e^{\lambda(\mu+t)}\right) \leq e^{-\lambda(\mu+t)}\, \mathbb{E}\left(e^{\lambda S_n}\right) = e^{-\lambda t} \prod_{i=1}^n \mathbb{E}\left(e^{\lambda(X_i - \mu_i)}\right).$$

Note that $e^{\lambda x}$ is a convex function of x, and since $0 \leq X_i \leq c_i$, we have

$$e^{\lambda(X_i-\mu_i)} \leq e^{-\lambda\mu_i}\left(1 - \frac{X_i}{c_i} + \frac{X_i}{c_i}e^{\lambda c_i}\right)$$

and so

$$\begin{aligned}\mathbb{E}(e^{\lambda(X_i-\mu_i)}) &\leq e^{-\lambda\mu_i}\left(1 - \frac{\mu_i}{c_i} + \frac{\mu_i}{c_i}e^{\lambda c_i}\right)\\ &= e^{-\theta_i p_i}\left(1 - p_i + p_i e^{\theta_i}\right),\end{aligned} \tag{2.36}$$

where $\theta_i = \lambda c_i$ and $p_i = \mu_i / c_i$.

Then, taking the logarithm of the RHS of (2.36), we have

$$\begin{aligned}f(\theta_i) &= -\theta_i p_i + \log\left(1 - p_i + p_i e^{\theta_i}\right),\\ f'(\theta_i) &= -p_i + \frac{p_i e^{\theta_i}}{1 - p_i + p_i e^{\theta_i}},\\ f''(\theta_i) &= \frac{p_i(1-p_i)e^{-\theta_i}}{((1-p_i)e^{-\theta_i} + p_i)^2}.\end{aligned}$$

Now $\frac{\alpha\beta}{(\alpha+\beta)^2} \leq 1/4$ and so $f''(\theta_i) \leq 1/4$, and therefore

$$f(\theta_i) \leq f(0) + f'(0)\theta_i + \frac{1}{8}\theta_i^2 = \frac{\lambda^2 c_i^2}{8}.$$

It follows then that

$$\mathbb{P}(S_n \geq \mu + t) \leq e^{-\lambda t} \exp\left\{\sum_{i=1}^{n} \frac{\lambda^2 c_i^2}{8}\right\}.$$

We obtain (2.34) by putting $\lambda = \frac{4}{\sum_{i=1}^n c_i^2}$, and (2.35) is proved in a similar manner. □

There are many cases when we want to use our inequalities to bound the upper tail of some random variable Y and (i) Y does not satisfy the necessary conditions to apply the relevant inequality, but (ii) Y is dominated by some random variable X that does.

We say that a random variable X *stochastically dominates* a random variable Y and write $X > Y$ if

$$\mathbb{P}(X \geq t) \geq \mathbb{P}(Y \geq t) \quad \text{for all real } t. \tag{2.37}$$

Clearly, we can use X as a surrogate for Y if (2.37) holds.

The following case arises quite often. Suppose that $Y = Y_1 + Y_2 + \cdots + Y_n$, where $Y_1, Y_2, \ldots, Y_n$ are not independent, but instead we have that for all t in the range $[A_i, B_i]$ of Y_i,

$$\mathbb{P}(Y_i \geq t \mid Y_1, Y_2, \ldots, Y_{i-1}) \leq \varphi(t),$$

where $\varphi(t)$ decreases monotonically from 1 to 0 in $[A_i, B_i]$.

Let X_i be a random variable taking values in the same range as Y_i and such that $\mathbb{P}(X_i \geq t) = \varphi(t)$. Let $X = X_1 + \cdots + X_n$, where $X_1, X_2, \ldots, X_n$ are independent of each other and $Y_1, Y_2, \ldots, Y_n$. Then we have

Lemma 2.29 *X stochastically dominates Y.*

Proof Let $X^{(i)} = X_1 + \cdots + X_i$ and $Y^{(i)} = Y_1 + \cdots + Y_i$ for $i = 1, 2, \ldots, n$. We will show by induction that $X^{(i)}$ dominates $Y^{(i)}$ for $i = 1, 2, \ldots, n$. This is trivially true for $i = 1$, and for $i > 1$ we have

$$\begin{aligned}\mathbb{P}(Y^{(i)} \geq t \mid Y_1, \ldots, Y_{i-1}) &= \mathbb{P}(Y_i \geq t - (Y_1 + \cdots + Y_{i-1}) \mid Y_1, \ldots, Y_{i-1}) \\ &\leq \mathbb{P}(X_i \geq t - (Y_1 + \cdots + Y_{i-1}) \mid Y_1, \ldots, Y_{i-1}).\end{aligned}$$

Removing the conditioning, we have

$$\mathbb{P}(Y^{(i)} \geq t) \leq \mathbb{P}(Y^{(i-1)} \geq t - X_i) \leq \mathbb{P}(X^{(i-1)} \geq t - X_i) = \mathbb{P}(X^{(i)} \geq t),$$

where the second inequality follows by induction. □

Exercises

2.3.1. Suppose we roll a fair die n times. Show that w.h.p. the number of odd outcomes is within $O(n^{1/2} \log n)$ of the number of even outcomes.

2.3.2. Consider the outcome of tossing a fair coin n times. Represent this by a (random) string of H's and T's. Show that w.h.p. there are $\sim n/8$ occurrences of HTH as a contiguous substring.

2.3.3. Check that (2.25) remains true for $x > 1$.
(Hint: differentiate both sides, twice.)

Problems for Chapter 2

2.1 Show that if $k = o(n)$, then

$$\binom{n}{k} \sim \left(\frac{ne}{k}\right)^k (2\pi k)^{-1/2} \exp\left\{-\frac{k^2}{2n}(1 + o(1))\right\}.$$

2.2 Let c be a constant, $0 < c < 1$, and let $k \sim cn$. Show that for such k,

$$\binom{n}{k} = 2^{n(H(c)+o(1))},$$

where H is an *entropy function*: $H(c) = -c \ln c - (1 - c) \ln(1 - c)$.

2.3 Prove the following strengthening of (2.2),

$$\sum_{\ell=0}^{k} \binom{n}{\ell} \leq \left(\frac{ne}{k}\right)^k.$$

2.4 Let $f(n) = \sum_{k=1}^{n} \frac{1}{k} \prod_{j=0}^{k-1} \left(1 - \frac{j}{n}\right)$. Prove that $f(n) \sim \frac{1}{2} \log n$.

2.5 Suppose that $m = cn$ distinguishable balls are thrown randomly into n boxes. (i) Write down an expression for the expected number of boxes that contain k or more balls. (ii) Show that your expression tends to zero if $k = \lceil \log n \rceil$.

2.6 Suppose that $m = cn$ distinguishable balls are thrown randomly into n boxes. Suppose that box i contains b_i balls. (i) Write down an expression for the expected number of k-sequences such that $b_i = b_{i+1} = \cdots = b_{i+k-1} = 0$. (ii) Show that your expression tends to zero if $k = \lceil \log n \rceil$.

2.7 Suppose that we toss a fair coin. Estimate the probability that we have to make $(2 + \varepsilon)n$ tosses before we see n heads.

2.8 Let $X_1, X_2, \ldots, X_n$ be independent binary random variables, $X_i \in \{0, 1\}$, and let $\mathbb{P}(X_i = 1) = p$ for every $1 \leq i \leq n$, where $0 < p < 1$. Let $\overline{S}_n = \frac{1}{n} \sum_{i=1}^{n} X_i$. Apply the Chernoff–Hoeffding bounds to show that if $n \geq (3/t^2) \ln(2/\delta)$, then $\mathbb{P}(|\overline{S}_n - p)| \leq t) \geq 1 - \delta$.

2.9 Let $Y_1, Y_2, \ldots, Y_m$ be independent non-negative integer random variables. Suppose that for $r \geq 1$ we have $\Pr(Y_r \geq k) \leq C\rho^k$, where $\rho < 1$. Let $\mu = C/(1-\rho)$. Show that if $Y = Y_1 + Y_2 + \cdots + Y_m$, then

$$\Pr(Y \geq (1 + \varepsilon)\mu m) \leq e^{-B\varepsilon^2 m}$$

for $0 \leq \varepsilon \leq 1$ and some $B = B(C, \rho)$.

2.10 We say that a sequence of random variables $A_0, A_1, \ldots$ is (η, N)-bounded if $A_i - \eta \leq A_{i+1} \leq A_i + N$ for all $i \geq 0$.
(i) Suppose that $\eta \leq N/2$ and $a < \eta m$. Prove that if $0 = A_0, A_1, \ldots$ is an (η, N)-bounded sequence, then $\Pr(A_m \leq -a) \leq \exp\left\{-\frac{a^2}{3\eta m N}\right\}$.
(ii) Suppose that $\eta \leq N/10$ and $a < \eta m$. Prove that if $0 = A_0, A_1, \ldots$ is an (η, N)-bounded sequence, then $\Pr(A_m \geq a) \leq \exp\left\{-\frac{a^2}{3\eta m N}\right\}$.

2.11 Let A be an $n \times m$ matrix, with each $a_{ij} \in \{0, 1\}$, and let $\vec{b}$ be an m-dimensional vector, with each $b_k \in \{-1, 1\}$, where each possibility is chosen with probability $1/2$. Let $\vec{c}$ be the n-dimensional vector that denotes the product of A and $\vec{b}$. Applying the Chernoff–Hoeffding bound show that the following inequality holds for $i \in \{1, 2, \ldots, n\}$:

$$\mathbb{P}(\max\{|c_i|\} \geq \sqrt{4m \ln n}) \leq O(n^{-1}).$$

Part II

Erdős–Rényi–Gilbert Model

3 Uniform and Binomial Random Graphs

There are two classic ways to generate a random graph. The first, introduced by Erdős and Rényi, involves sampling, uniformly at random, a single graph from the family of all labeled graphs on the vertex set $[n]$ with m edges. This is equivalent to the insertion of m randomly chosen edges into an empty graph on n vertices. Each choice of m places among the $\binom{n}{2}$ possibilities is equally likely. Gilbert suggested an alternative approach, where each edge is inserted into an empty graph on n vertices, independently and with the same probability p. One may immediately notice the main difference between those two approaches: the first one has a fixed number of edges m, while the number of edges in the second one is not fixed but random! Regardless of this fundamental difference, it appears that those two models are, in the probabilistic and asymptotic sense, equivalent when the number of edges m in the uniform model is approximately equal to the expected number of edges in the latter one, i.e., in such circumstances both models are almost indistinguishable. This is the reason why we think about them as a single, unified Erdős–Rényi–Gilbert model. In this chapter we formally introduce both Erdős–Rényi and Gilbert models, study their relationships and establish conditions for their asymptotic equivalence. We also define and study the basic features of the asymptotic behavior of random graphs, i.e., the existence of thresholds for monotone graph properties.

3.1 Models and Relationships

The study of random graphs in their own right began in earnest with the seminal paper of Erdős and Rényi [43]. This paper was the first to exhibit the threshold phenomena that characterize the subject.

Let $\mathcal{G}_{n,m}$ be the family of all labeled graphs with vertex set $V = [n] = \{1, 2, \dots, n\}$ and exactly m edges, $0 \le m \le \binom{n}{2}$. To every graph $G \in \mathcal{G}_{n,m}$, we assign a probability

$$\mathbb{P}(G) = \binom{\binom{n}{2}}{m}^{-1}.$$

Equivalently, we start with an empty graph on the set $[n]$ and insert m edges in such a way that all possible $\binom{\binom{n}{2}}{m}$ choices are equally likely. We denote such a random graph by $\mathbb{G}_{n,m} = ([n], E_{n,m})$ and call it a *uniform random graph*. We now describe a similar model. Fix $0 \le p = p(n) \le 1$. Then for $0 \le m \le \binom{n}{2}$, assign to each graph G with vertex set $[n]$ and m edges a probability

$$\mathbb{P}(G) = p^m (1-p)^{\binom{n}{2}-m}.$$

Equivalently, we start with an empty graph with vertex set $[n]$ and perform $\binom{n}{2}$ Bernoulli experiments inserting edges independently with probability p. We call such a random graph a *binomial random graph* and denote it by $\mathbb{G}_{n,p} = ([n], E_{n,p})$. This model was introduced by Gilbert [54]. As one may expect, there is a close relationship between these two models of random graphs. We start with a simple observation.

Lemma 3.1 *The random graph $\mathbb{G}_{n,p}$, given that its number of edges is m, is equally likely to be one of the $\binom{\binom{n}{2}}{m}$ graphs that have m edges.*

Proof Let G_0 be any labeled graph with m edges. Then since

$$\{\mathbb{G}_{n,p} = G_0\} \subseteq \{|E_{n,p}| = m\},$$

we have

$$\begin{aligned}
\mathbb{P}(\mathbb{G}_{n,p} = G_0 \mid |E_{n,p}| = m) &= \frac{\mathbb{P}(\mathbb{G}_{n,p} = G_0, |E_{n,p}| = m)}{\mathbb{P}(|E_{n,p}| = m)} \\
&= \frac{\mathbb{P}(\mathbb{G}_{n,p} = G_0)}{\mathbb{P}(|E_{n,p}| = m)} \\
&= \frac{p^m (1-p)^{\binom{n}{2}-m}}{\binom{\binom{n}{2}}{m} p^m (1-p)^{\binom{n}{2}-m}} \\
&= \binom{\binom{n}{2}}{m}^{-1}.
\end{aligned}$$

□

Thus $\mathbb{G}_{n,p}$ conditioned on the event $\{\mathbb{G}_{n,p}$ has m edges$\}$ is equal in distribution to $\mathbb{G}_{n,m}$, the graph chosen uniformly at random from all graphs with m edges.

Obviously, the main difference between those two models of random graphs is that in $\mathbb{G}_{n,m}$ we choose its number of edges, while in the case of $\mathbb{G}_{n,p}$ the number of edges is the binomial random variable with the parameters $\binom{n}{2}$ and p. Intuitively, for large n random graphs $\mathbb{G}_{n,m}$ and $\mathbb{G}_{n,p}$ should behave in a similar fashion when the number of edges m in $\mathbb{G}_{n,m}$ equals or is "close" to the expected number of edges of $\mathbb{G}_{n,p}$, i.e., when

$$m = \binom{n}{2} p \sim \frac{n^2 p}{2}, \tag{3.1}$$

or, equivalently, when the edge probability in $\mathbb{G}_{n,p}$

$$p \sim \frac{2m}{n^2}. \tag{3.2}$$

We next introduce a useful "coupling technique" that generates the random graph

$\mathbb{G}_{n,p}$ in two independent steps. We will then describe a similar idea in relation to $\mathbb{G}_{n,m}$.

Let $\mathbb{G}_{n,p}$ be a union of two independent random graphs $\mathbb{G}_{n,p_1}$ and $\mathbb{G}_{n,p_2}$, i.e.,

$$\mathbb{G}_{n,p} = \mathbb{G}_{n,p_1} \cup \mathbb{G}_{n,p_2}.$$

Suppose that $p_1 < p$ and p_2 is defined by the equation

$$1 - p = (1 - p_1)(1 - p_2), \tag{3.3}$$

or, equivalently,

$$p = p_1 + p_2 - p_1 p_2.$$

Thus an edge is not included in $\mathbb{G}_{n,p}$ if it is not included in either of $\mathbb{G}_{n,p_1}$ or $\mathbb{G}_{n,p_2}$. So when we write

$$\mathbb{G}_{n,p_1} \subseteq \mathbb{G}_{n,p},$$

we mean that the two graphs are *coupled* so that $\mathbb{G}_{n,p}$ is obtained from $\mathbb{G}_{n,p_1}$ by superimposing it with $\mathbb{G}_{n,p_2}$ and replacing any double edges by a single one.

We can also couple random graphs $\mathbb{G}_{n,m_1}$ and $\mathbb{G}_{n,m_2}$, where $m_2 \geq m_1$, via

$$\mathbb{G}_{n,m_2} = \mathbb{G}_{n,m_1} \cup \mathbb{H}.$$

Here $\mathbb{H}$ is the random graph on vertex set $[n]$ that has $m_2 - m_1$ edges chosen uniformly at random from $\binom{[n]}{2} \setminus E_{n,m_1}$.

Consider now a graph property $\mathcal{P}$ defined as a subset of the set of all labeled graphs on vertex set $[n]$, i.e., $\mathcal{P} \subseteq 2^{\binom{n}{2}}$. For example, all connected graphs (on n vertices), graphs with a Hamiltonian cycle, graphs containing a given subgraph, planar graphs, and graphs with a vertex of given degree form a specific "graph property."

We will state below two simple observations which show a general relationship between $\mathbb{G}_{n,m}$ and $\mathbb{G}_{n,p}$ in the context of the probabilities of having a given graph property $\mathcal{P}$. The constant 10 in the next lemma is not best possible, but in the context of the usage of the lemma, any constant will suffice.

Lemma 3.2 *Let $\mathcal{P}$ be any graph property and $p = m/N$, $N = \binom{n}{2}$, where $m = m(n) \to \infty$, $N - m \to \infty$. Then, for large n,*

$$\mathbb{P}(\mathbb{G}_{n,m} \in \mathcal{P}) \leq 10m^{1/2}\, \mathbb{P}(\mathbb{G}_{n,p} \in \mathcal{P}).$$

Proof By the law of total probability,

$$\begin{aligned} \mathbb{P}(\mathbb{G}_{n,p} \in \mathcal{P}) &= \sum_{k=0}^{N} \mathbb{P}(\mathbb{G}_{n,p} \in \mathcal{P} \mid |E_{n,p}| = k)\, \mathbb{P}(|E_{n,p}| = k) \\ &= \sum_{k=0}^{N} \mathbb{P}(\mathbb{G}_{n,k} \in \mathcal{P})\, \mathbb{P}(|E_{n,p}| = k) \qquad (3.4) \\ &\geq \mathbb{P}(\mathbb{G}_{n,m} \in \mathcal{P})\, \mathbb{P}(|E_{n,p}| = m). \end{aligned}$$

To justify (3.4), we write

$$\mathbb{P}(\mathbb{G}_{n,p} \in \mathcal{P} \mid |E_{n,p}| = k) = \frac{\mathbb{P}(\mathbb{G}_{n,p} \in \mathcal{P} \wedge |E_{n,p}| = k)}{\mathbb{P}(|E_{n,p}| = k)}$$

$$= \sum_{\substack{G \in \mathcal{P} \\ |E(G)|=k}} \frac{p^k (1-p)^{N-k}}{\binom{N}{k} p^k (1-p)^{N-k}}$$
$$= \sum_{\substack{G \in \mathcal{P} \\ |E(G)|=k}} \frac{1}{\binom{N}{k}}$$
$$= \mathbb{P}(\mathbb{G}_{n,k} \in \mathcal{P}).$$

Next recall that the number of edges $|E_{n,p}|$ of a random graph $\mathbb{G}_{n,p}$ is a random variable with the binomial distribution with parameters $\binom{n}{2}$ and p. Applying Stirling's formula (see Lemma 2.4) for the factorials in $\binom{N}{m}$, we get

$$\begin{aligned}
\mathbb{P}(|E_{n,p}| = m) &= \binom{N}{m} p^m (1-p)^{\binom{n}{2}-m} \\
&= (1+o(1)) \frac{N^N \sqrt{2\pi N}\, p^m (1-p)^{N-m}}{m^m (N-m)^{N-m}\, 2\pi \sqrt{m(N-m)}} \qquad (3.5) \\
&= (1+o(1)) \sqrt{\frac{N}{2\pi m(N-m)}}.
\end{aligned}$$

Hence

$$\mathbb{P}(|E_{n,p}| = m) \geq \frac{1}{10\sqrt{m}},$$

and

$$\mathbb{P}(\mathbb{G}_{n,m} \in \mathcal{P}) \leq 10 m^{1/2}\, \mathbb{P}(\mathbb{G}_{n,p} \in \mathcal{P}).$$

□

We call a graph property $\mathcal{P}$ *monotone increasing* if $G \in \mathcal{P}$ implies $G + e \in \mathcal{P}$, i.e., adding an edge e to a graph G does not destroy the property.

A monotone increasing property is *nontrivial* if the empty graph $\bar{K}_n \notin \mathcal{P}$ and the complete graph $K_n \in \mathcal{P}$.

A graph property is *monotone decreasing* if $G \in \mathcal{P}$ implies $G - e \in \mathcal{P}$, i.e., removing an edge from a graph does not destroy the property.

For example, connectivity and Hamiltonicity are monotone increasing properties, while the properties of a graph not being connected or being planar are examples of monotone decreasing graph properties. Obviously, a graph property $\mathcal{P}$ is monotone increasing if and only if its complement is monotone decreasing. Clearly, not all graph properties are monotone. For example, having at least half of the vertices having a given fixed degree d is not monotone.

From the coupling argument it follows that if $\mathcal{P}$ is a monotone increasing property, then, whenever $p < p'$ or $m < m'$,

$$\mathbb{P}(\mathbb{G}_{n,p} \in \mathcal{P}) \leq \mathbb{P}(\mathbb{G}_{n,p'} \in \mathcal{P}) \qquad (3.6)$$

and

$$\mathbb{P}(\mathbb{G}_{n,m} \in \mathcal{P}) \leq \mathbb{P}(\mathbb{G}_{n,m'} \in \mathcal{P}), \tag{3.7}$$

respectively.

For monotone increasing graph properties we can get a much better upper bound on $\mathbb{P}(\mathbb{G}_{n,m} \in \mathcal{P})$, in terms of $\mathbb{P}(\mathbb{G}_{n,p} \in \mathcal{P})$, than that given by Lemma 3.2.

Lemma 3.3 *Let $\mathcal{P}$ be a monotone increasing graph property and $p = m/N$, $N = \binom{n}{2}$. Then, for large n and $p = o(1)$ such that $Np, N(1-p)/(Np)^{1/2} \to \infty$,*

$$\mathbb{P}(\mathbb{G}_{n,m} \in \mathcal{P}) \leq 3\,\mathbb{P}(\mathbb{G}_{n,p} \in \mathcal{P}).$$

Proof Suppose $\mathcal{P}$ is monotone increasing and $p = m/N$. Then

$$\begin{aligned}\mathbb{P}(\mathbb{G}_{n,p} \in \mathcal{P}) &= \sum_{k=0}^{N} \mathbb{P}(\mathbb{G}_{n,k} \in \mathcal{P})\,\mathbb{P}(|E_{n,p}| = k)\\ &\geq \sum_{k=m}^{N} \mathbb{P}(\mathbb{G}_{n,k} \in \mathcal{P})\,\mathbb{P}(|E_{n,p}| = k).\end{aligned}$$

However, by the coupling property we know that for $k \geq m$,

$$\mathbb{P}(\mathbb{G}_{n,k} \in \mathcal{P}) \geq \mathbb{P}(\mathbb{G}_{n,m} \in \mathcal{P}).$$

The number of edges $|E_{n,p}|$ in $\mathbb{G}_{n,p}$ has the binomial distribution with parameters N, p. Hence

$$\begin{aligned}\mathbb{P}(\mathbb{G}_{n,p} \in \mathcal{P}) &\geq \mathbb{P}(\mathbb{G}_{n,m} \in \mathcal{P}) \sum_{k=m}^{N} \mathbb{P}(|E_{n,p}| = k)\\ &= \mathbb{P}(\mathbb{G}_{n,m} \in \mathcal{P}) \sum_{k=m}^{N} u_k,\end{aligned} \tag{3.8}$$

where

$$u_k = \binom{N}{k} p^k (1-p)^{N-k}.$$

Now, using Stirling's approximation (2.4),

$$u_m = (1+o(1))\frac{N^N p^m (1-p)^{N-m}}{m^m (N-m)^{N-m} (2\pi m)^{1/2}} = \frac{1+o(1)}{(2\pi m)^{1/2}}.$$

Furthermore, if $k = m + t$ where $0 \leq t \leq m^{1/2}$, then

$$\frac{u_{k+1}}{u_k} = \frac{(N-k)p}{(k+1)(1-p)} = \frac{1 - \frac{t}{N-m}}{1 + \frac{t+1}{m}} \geq \exp\left\{-\frac{t}{N-m-t} - \frac{t+1}{m}\right\},$$

after using the following bounds:

$$1 + x \le e^x \text{ for every } x, \tag{3.9}$$

$$1 - x \ge e^{-x/(1-x)} \text{ for } 0 \le x \le 1 \tag{3.10}$$

to obtain the inequality and our assumptions on N, p.

It follows that for $0 \le t \le m^{1/2}$,

$$u_{m+t} \ge \frac{1+o(1)}{(2\pi m)^{1/2}} \exp\left\{-\sum_{s=0}^{t-1}\left(\frac{s}{N-m-s} - \frac{s+1}{m}\right)\right\} \ge \frac{\exp\left\{-\frac{t^2}{2m} - o(1)\right\}}{(2\pi m)^{1/2}},$$

where we have used the fact that $m = o(N)$.

It follows that

$$\sum_{k=m}^{m+m^{1/2}} u_k \ge \frac{1-o(1)}{(2\pi)^{1/2}} \int_{x=0}^{1} e^{-x^2/2} dx \ge \frac{1}{3},$$

and the lemma follows from (3.8). □

Lemmas 3.2 and 3.3 are surprisingly applicable. In fact, since the $\mathbb{G}_{n,p}$ model is computationally easier to handle than $\mathbb{G}_{n,m}$, we will repeatedly use both lemmas to show that $\mathbb{P}(\mathbb{G}_{n,p} \in \mathcal{P}) \to 0$ implies that $\mathbb{P}(\mathbb{G}_{n,m} \in \mathcal{P}) \to 0$ when $n \to \infty$. In other situations we can use a stronger and more widely applicable result. The theorem below, which we state without proof, gives precise conditions for the asymptotic equivalence of random graphs $\mathbb{G}_{n,p}$ and $\mathbb{G}_{n,m}$. It is due to Łuczak [79].

Theorem 3.4 *Let $0 \le p_0 \le 1$, $s(n) = n\sqrt{p(1-p)} \to \infty$, and $\omega(n) \to \infty$ arbitrarily slowly as $n \to \infty$.*

(i) Suppose that $\mathcal{P}$ is a graph property such that $\mathbb{P}(\mathbb{G}_{n,m} \in \mathcal{P}) \to p_0$ for all

$$m \in \left[\binom{n}{2}p - \omega(n)s(n), \binom{n}{2}p + \omega(n)s(n)\right].$$

Then $\mathbb{P}(\mathbb{G}_{n,p} \in \mathcal{P}) \to p_0$ as $n \to \infty$.

(ii) Let $p_- = p - \omega(n)s(n)/n^2$ and $p_+ = p + \omega(n)s(n)/n^2$. Suppose that $\mathcal{P}$ is a monotone graph property such that $\mathbb{P}(\mathbb{G}_{n,p_-} \in \mathcal{P}) \to p_0$ and $\mathbb{P}(\mathbb{G}_{n,p_+} \in \mathcal{P}) \to p_0$. Then $\mathbb{P}(\mathbb{G}_{n,m} \in \mathcal{P}) \to p_0$ as $n \to \infty$, where $m = \lfloor \binom{n}{2} p \rfloor$.

Exercises

3.1.1. Compute the expected number of triangles in $\mathbb{G}_{n,p}$ and $\mathbb{G}_{n,m}$ and show when these parameters are asymptotically equal. Compute also the variance of both random variables.

3.1.2. Compute the expected number of copies of K_4 (the complete graph on four vertices) and $\mathbb{G}_{n,m}$ and show when those parameters are asymptotically equal.

3.1.3. Consider a graph property $\mathcal{P}$ defined as a subset of the set of all graphs on n vertices. Are the following graph properties monotone increasing, monotone decreasing or nonmonotone?
$\mathcal{P} = \{G : s.t.\ G \text{ contains an isolated vertex}\}$
$\mathcal{P} = \{G : s.t.\ G \text{ contains a subgraph } H\}$
$\mathcal{P} = \{G : s.t.\ G \text{ contains an induced subgraph } H\}$
$\mathcal{P} = \{G : s.t.\ G \text{ has a perfect matching}\}$
$\mathcal{P} = \{G : s.t. \text{ the largest component of } G \text{ is a tree}\}$
$\mathcal{P} = \{G : s.t. \text{ all vertex degrees are at most } \Delta\}$
$\mathcal{P} = \{G : s.t.\ G \text{ has a chromatic number equal to } 3\}$
$\mathcal{P} = \{G : s.t.\ G \text{ has at least } k \text{ vertices of given degree}\}$
$\mathcal{P} = \{G : s.t.\ G \text{ is nonplanar}\}$

3.1.4. Construct a few of your own examples of monotone and nonmonotone properties.

3.1.5. Prove that graph property $\mathcal{P}$ is increasing if and only if its complement $\mathcal{P}^c$ is decreasing.

3.1.6. Prove (3.9) and (3.10).

3.1.7. Prove that $\binom{N}{k}p^k(1-p)^{N-k} = (1+o(1))\sqrt{\frac{N}{2\pi k(N-k)}}$, where $N = \binom{n}{2}$ and $p = \frac{k}{N}$.

3.2 Thresholds

One of the most striking observations regarding the asymptotic properties of random graphs is the "abrupt" nature of the appearance and disappearance of certain graph properties. To be more precise in the description of this phenomenon, let us introduce *threshold functions* (or just *thresholds*) for monotone graph properties. We start by giving the formal definition of a threshold for a monotone increasing graph property $\mathcal{P}$.

Definition 3.5 A function $m^* = m^*(n)$ is a *threshold* for a monotone increasing property $\mathcal{P}$ in the random graph $\mathbb{G}_{n,m}$ if

$$\lim_{n\to\infty} \mathbb{P}(\mathbb{G}_{n,m} \in \mathcal{P}) = \begin{cases} 0 & \text{if } m/m^* \to 0, \\ 1 & \text{if } m/m^* \to \infty, \end{cases}$$

as $n \to \infty$.

A similar definition applies to the edge probability $p = p(n)$ in a random graph $\mathbb{G}_{n,p}$.

Definition 3.6 A function $p^* = p^*(n)$ is a *threshold* for a monotone increasing property $\mathcal{P}$ in the random graph $\mathbb{G}_{n,p}$ if

$$\lim_{n\to\infty} \mathbb{P}(\mathbb{G}_{n,p} \in \mathcal{P}) = \begin{cases} 0 & \text{if } p/p^* \to 0, \\ 1 & \text{if } p/p^* \to \infty, \end{cases}$$

as $n \to \infty$.

It is easy to see how to define thresholds for monotone decreasing graph properties, and therefore we will leave this to the reader.

Notice also that the thresholds defined above are not unique since any function which differs from $m^*(n)$ (resp. $p^*(n)$) by a constant factor is also a threshold for $\mathcal{P}$.

We will illustrate thresholds in a series of examples dealing with very simple graph properties. Our goal at the moment is to demonstrate some basic techniques to determine thresholds rather than to "discover" some "striking" facts about random graphs.

A standard way to show the first part of the threshold statement, i.e., that the probability that $\mathbb{G}_{n,m}$ (resp. $\mathbb{G}_{n,p}$) has property $\mathcal{P}$ tends to zero when $m \ll m^*$ (resp. $p \ll p^*$) as $n \to \infty$ is an application of the First Moment Method, which stems directly from the Markov inequality (see Lemma 2.15). Putting $t = 1$ in the Markov inequality we get:

First Moment Method If X is a non-negative integer-valued random variable, then

$$\mathbb{P}(X \geq 1) \leq \mathbb{E}X. \tag{3.11}$$

We start with the random graph $\mathbb{G}_{n,p}$ and the following properties:

$$\mathcal{P}_1 = \{\text{all nonempty (nonedgeless) labeled graphs on } n \text{ vertices}\}, \text{ and}$$

$$\mathcal{P}_2 = \{\text{all labeled graphs on } n \text{ vertices containing at least one triangle}\}.$$

Obviously, both graph properties are monotone increasing, and our goal will be to find thresholds for both of them.

Theorem 3.7 *$\mathbb{P}(\mathbb{G}_{n,p} \in \mathcal{P}_1) \to 0$ if $p \ll 1/n^2$, while $\mathbb{P}(\mathbb{G}_{n,p} \in \mathcal{P}_2) \to 0$ if $p \ll 1/n$, as $n \to \infty$.*

Proof Let X be a random variable counting the number of edges in $\mathbb{G}_{n,p}$. Then

$$\mathbb{P}(\mathbb{G}_{n,p} \in \mathcal{P}_1) = \mathbb{P}(\mathbb{G}_{n,p} \text{ has at least one edge}) = \mathbb{P}(X > 0).$$

Since X has the binomial distribution,

$$\mathbb{E}\,X = \binom{n}{2}p$$

and, by the First Moment Method,

$$\mathbb{P}(X > 0) \le \frac{n^2}{2} p \to 0$$

as $n \to \infty$, when $p \ll n^{-2}$.

Similarly, let Z be the number of triangles in $\mathbb{G}_{n,p}$. Then

$$\mathbb{E}\, Z = \binom{n}{3} p^3. \tag{3.12}$$

We can see this as follows. Let $T_1, T_2, \dots, T_{\binom{n}{3}}$ be an enumeration of the triangles of the complete graph K_n. Also, let Z_i be the indicator for $G_{n,p}$ to contain the triangle T_i. Then we have $\mathbb{E}\, Z_i = p^3$ for all i and $Z = Z_1 + \cdots + Z_{\binom{n}{3}}$ and (3.12) follows. Now, from (3.12) we get that $\mathbb{E}\, Z \to 0$ as $n \to \infty$, if $p \ll n^{-1}$. So the second statement also follows by the First Moment Method.

□

On the other hand, if we want to show that $\mathbb{P}(X > 0) \to 1$ (resp. $\mathbb{P}(Z > 0) \to 1$) as $n \to \infty$, then we cannot use the First Moment Method and we should apply the Second Moment Method, which is a simple consequence of the Chebyshev inequality.

Second Moment Method If X is a non-negative integer-valued random variable, then

$$\mathbb{P}(X \ge 1) \ge 1 - \frac{\operatorname{Var} X}{(\mathbb{E}\, X)^2}. \tag{3.13}$$

Proof Set $t = \mathbb{E}\, X$ in the Chebyshev inequality. Then

$$\mathbb{P}(X = 0) \le \mathbb{P}(|X - \mathbb{E}\, X| \ge \mathbb{E}\, X) \le \frac{\operatorname{Var} X}{(\mathbb{E}\, X)^2}.$$

□

(Strong) Second Moment Method If X is a non-negative integer-valued random variable, then

$$\mathbb{P}(X \ge 1) \ge \frac{(\mathbb{E}\, X)^2}{\mathbb{E}\, X^2}. \tag{3.14}$$

Proof Notice that

$$X = X \cdot I_{\{X \ge 1\}}.$$

Then, by the Cauchy–Schwarz inequality,

$$(\mathbb{E}\, X)^2 = \left(\mathbb{E}(X \cdot I_{\{X \ge 1\}})\right)^2 \le \mathbb{E}\, I^2_{\{X \ge 1\}}\, \mathbb{E}\, X^2 = \mathbb{P}(X \ge 1)\, \mathbb{E}\, X^2.$$

□

Let us complete our discussion about thresholds for properties $\mathcal{P}_1$ and $\mathcal{P}_2$.

Theorem 3.8 *$\mathbb{P}(\mathbb{G}_{n,p} \in \mathcal{P}_1) \to 1$ if $p \gg 1/n^2$, while $\mathbb{P}(\mathbb{G}_{n,p} \in \mathcal{P}_2) \to 1$ if $p \gg 1/n$ as $n \to \infty$.*

Proof Recall that the random variable X denotes the number of edges in the random graph $\mathbb{G}_{n,p}$ and has the binomial distribution. Therefore

$$\operatorname{Var} X = \binom{n}{2} p(1-p) = (1-p)\,\mathbb{E}\,X.$$

By the Second Moment Method, $\mathbb{P}(X \geq 1) \to 1$ as $n \to \infty$ whenever $\operatorname{Var} X/(\mathbb{E}\,X)^2 \to 0$ as $n \to \infty$. Now, if $p \gg n^{-2}$, then $\mathbb{E}\,X \to \infty$, and therefore

$$\frac{\operatorname{Var} X}{(\mathbb{E}\,X)^2} = \frac{1-p}{\mathbb{E}\,X} \to 0$$

as $n \to \infty$, which shows that indeed $\mathbb{P}(\mathbb{G}_{n,p} \in \mathcal{P}_1) \to 1$ if $p \gg 1/n^2$ as $n \to \infty$.

To show that if $np \to \infty$ then $\mathbb{P}(\mathbb{G}_{n,p}$ contains at least one triangle) as $n \to \infty$ needs a bit more work.

Assume first that $np = \omega \leq \log n$, where $\omega = \omega(n) \to \infty$. Let Z, as before, denote the number of triangles in $\mathbb{G}_{n,p}$. Then

$$\mathbb{E}\,Z = \binom{n}{3} p^3 \geq (1 - o(1))\frac{\omega^3}{6} \to \infty.$$

We remind the reader that simply having $\mathbb{E}\,Z \to \infty$ is not sufficient to prove that $\mathbb{P}(Z > 0) \to 1$.

Next let $T_1, T_2, \ldots, T_M$, $M = \binom{n}{3}$ be an enumeration of the triangles of K_n. Then

$$\begin{aligned}
\mathbb{E}\,Z^2 &= \sum_{i,j=1}^{M} \mathbb{P}(T_i, T_j \in \mathbb{G}_{n,p}) \\
&= \sum_{i=1}^{M} \mathbb{P}(T_i \in \mathbb{G}_{n,p}) \sum_{j=1}^{M} \mathbb{P}(T_j \in \mathbb{G}_{n,p} \mid T_i \in \mathbb{G}_{n,p}) \qquad (3.15) \\
&= M\,\mathbb{P}(T_1 \in \mathbb{G}_{n,p}) \sum_{j=1}^{M} \mathbb{P}(T_j \in \mathbb{G}_{n,p} \mid T_1 \in \mathbb{G}_{n,p}) \qquad (3.16) \\
&= \mathbb{E}\,Z \times \sum_{j=1}^{M} \mathbb{P}(T_j \in \mathbb{G}_{n,p} \mid T_1 \in \mathbb{G}_{n,p}).
\end{aligned}$$

Here (3.16) follows from (3.15) by symmetry.

Now suppose that T_j, T_1 share σ_j edges. Then

$$\begin{aligned}
&\sum_{j=1}^{M} \mathbb{P}(T_j \in \mathbb{G}_{n,p} \mid T_1 \in \mathbb{G}_{n,p}) \\
&\qquad = \sum_{j:\sigma_j=3} \mathbb{P}(T_j \in \mathbb{G}_{n,p} \mid T_1 \in \mathbb{G}_{n,p})
\end{aligned}$$

$$+\sum_{j:\sigma_j=1} \mathbb{P}(T_j \in \mathbb{G}_{n,p} \mid T_1 \in \mathbb{G}_{n,p})$$
$$+\sum_{j:\sigma_j=0} \mathbb{P}(T_j \in \mathbb{G}_{n,p} \mid T_1 \in \mathbb{G}_{n,p})$$
$$= 1 + 3(n-3)p^2 + \left(\binom{n}{3} - 3n + 8\right) p^3$$
$$\leq 1 + \frac{3\omega^2}{n} + \mathbb{E}\, Z.$$

It follows that

$$\operatorname{Var} Z \leq (\mathbb{E}\, Z)\left(1 + \frac{3\omega^2}{n} + \mathbb{E}\, Z\right) - (\mathbb{E}\, Z)^2 \leq 2\,\mathbb{E}\, Z.$$

Applying the Second Moment Method we get

$$\mathbb{P}(Z = 0) \leq \frac{\operatorname{Var} Z}{(\mathbb{E}\, Z)^2} \leq \frac{2}{\mathbb{E}\, Z} = o(1).$$

This proves the statement for $p \leq \frac{\log n}{n}$. For larger p we can use (3.6). □

Summarizing the results of both examples we see that $p^* = n^{-2}$ is the threshold for the property that a random graph $\mathbb{G}_{n,p}$ contains at least one edge (is nonempty), while $p^* = n^{-1}$ is the threshold for the property that it contains at least one triangle (is not triangle free).

Consider the monotone decreasing graph property that a graph contains an isolated vertex, i.e., a vertex of degree zero:

$$\mathcal{P} = \{\text{all labeled graphs on } n \text{ vertices containing isolated vertices}\}.$$

We will show that $m^* = \frac{1}{2} n \log n$ is a threshold function for the above property $\mathcal{P}$ in $\mathbb{G}_{n,m}$.

Theorem 3.9 *Let $\mathcal{P}$ be the property that a graph on n vertices contains at least one isolated vertex and let $m = \frac{1}{2} n(\log n + \omega(n))$. Then*

$$\lim_{n\to\infty} \mathbb{P}(\mathbb{G}_{n,m} \in \mathcal{P}) = \begin{cases} 1 & \text{if } \omega(n) \to -\infty, \\ 0 & \text{if } \omega(n) \to \infty. \end{cases}$$

To see that the second statement holds we use the First Moment Method. Namely, let $X_0 = X_{n,0}$ be the number of isolated vertices in the random graph $\mathbb{G}_{n,m}$. Then X_0 can be represented as the sum of indicator random variables

$$X_0 = \sum_{v \in V} I_v,$$

where

$$I_v = \begin{cases} 1 & \text{if } v \text{ is an isolated vertex in } \mathbb{G}_{n,m}, \\ 0 & \text{otherwise.} \end{cases}$$

So

$$\mathbb{E}\,X_0 = \sum_{v\in V} \mathbb{E}\,I_v = n\frac{\binom{\binom{n-1}{2}}{m}}{\binom{\binom{n}{2}}{m}} = n\left(\frac{n-2}{n}\right)^m \prod_{i=0}^{m-1}\left(1 - \frac{4i}{n(n-1)(n-2)-2i(n-2)}\right)$$
$$= n\left(\frac{n-2}{n}\right)^m \left(1 + O\left(\frac{(\log n)^2}{n}\right)\right) \tag{3.17}$$

using (2.5) and assuming that $\omega = o(\log n)$, while for the product we use (2.6).

Hence, by (3.9),

$$\mathbb{E}\,X_0 \le n\left(\frac{n-2}{n}\right)^m \le ne^{-2m/n} = e^{-\omega}$$

for $m = \frac{1}{2}n(\log n + \omega(n))$.

So $\mathbb{E}\,X_0 \to 0$ when $\omega(n) \to \infty$ as $n \to \infty$, and the First Moment Method implies that $X_0 = 0$ with probability tending to 1 as $n \to \infty$.

To show that the first statement holds in the case when $\omega \to -\infty$ we first observe from (3.17) that in this case

$$\mathbb{E}\,X_0 = (1-o(1))n\left(\frac{n-2}{n}\right)^m$$
$$\ge (1-o(1))n\exp\left\{-\frac{2m}{n-2}\right\}$$
$$\ge (1-o(1))e^{\omega} \to \infty. \tag{3.18}$$

The second inequality in the above comes from basic inequality (3.10), and we have once again assumed that $\omega = o(\log n)$ to justify the first equation.

We caution the reader that as before, $\mathbb{E}\,X_0 \to \infty$ does not prove that $\mathbb{P}(X_0 > 0) \to 1$ as $n \to \infty$. In Chapter 7 we will see an example of a random variable X_H, where $\mathbb{E}\,X_H \to \infty$ and yet $\mathbb{P}(X_H = 0) \to 1$ as $n \to \infty$.

Notice that

$$\mathbb{E}\,X_0^2 = \mathbb{E}\left(\sum_{v\in V} I_v\right)^2 = \sum_{u,v\in V} \mathbb{E}(I_u I_v)$$
$$= \sum_{u,v\in V} \mathbb{P}(I_u = 1, I_v = 1)$$
$$= \sum_{u\ne v} \mathbb{P}(I_u = 1, I_v = 1) + \sum_{u=v} \mathbb{P}(I_u = 1, I_v = 1)$$
$$= n(n-1)\frac{\binom{\binom{n-2}{2}}{m}}{\binom{\binom{n}{2}}{m}} + \mathbb{E}\,X_0$$

$$\leq n^2 \left(\frac{n-2}{n}\right)^{2m} + \mathbb{E}\, X_0$$
$$= (1 + o(1))(\mathbb{E}\, X_0)^2 + \mathbb{E}\, X_0.$$

The last equation follows from (3.17).

Hence, by the (strong) Second Moment Method,

$$\begin{aligned}\mathbb{P}(X_0 \geq 1) &\geq \frac{(\mathbb{E}\, X_0)^2}{\mathbb{E}\, X_0^2} \\ &= \frac{(\mathbb{E}\, X_0)^2}{(1+o(1))(\mathbb{E}\, X_0)^2 + \mathbb{E}\, X_0} \\ &= \frac{1}{(1+o(1)) + (\mathbb{E}\, X_0)^{-1}} \\ &= 1 - o(1)\end{aligned}$$

on using (3.18). Hence $\mathbb{P}(X_0 \geq 1) \to 1$ when $\omega(n) \to -\infty$ as $n \to \infty$, and so we can conclude that $m = m(n)$ is the threshold for the property that $\mathbb{G}_{n,m}$ contains isolated vertices. □

Note that the above result indicates that now the threshold m^* is more "sensitive" than that considered in Theorem 3.8, since the "switch" from probability one to probability zero appears if either $m/m^* \leq 1 - \varepsilon$ or $m/m^* \geq 1 + \varepsilon$. We will see later other situations where we can observe that for some monotone graph properties such more "sensitive" thresholds hold.

For this simple random variable X_0, we worked with $\mathbb{G}_{n,m}$. We will in general work with the more congenial independent model $\mathbb{G}_{n,p}$ and translate the results to $G_{n,m}$ if so desired.

A large body of the theory of random graphs is concerned with the search for thresholds for various properties, such as containing a path or cycle of a given length, or, in general, a copy of a given graph, or being connected or Hamiltonian, to name just a few. Therefore, the next result is of special importance. It was proved by Bollobás and Thomason [29].

Theorem 3.10 *Every nontrivial monotone graph property has a threshold.*

Proof Without loss of generality assume that $\mathcal{P}$ is a monotone increasing graph property. Given $0 < \varepsilon < 1$, we define $p(\varepsilon)$ by

$$\mathbb{P}(\mathbb{G}_{n,p(\varepsilon)} \in \mathcal{P}) = \varepsilon.$$

Note that $p(\varepsilon)$ exists because

$$\mathbb{P}(\mathbb{G}_{n,p} \in \mathcal{P}) = \sum_{G \in \mathcal{P}} p^{|E(G)|}(1-p)^{N-|E(G|}$$

is a polynomial in p that increases from 0 to 1. This is not obvious from the expression, but it is obvious from the fact that $\mathcal{P}$ is monotone increasing and that increasing p increases the likelihood that $\mathbb{G}_{n,p} \in \mathcal{P}$ (see (3.6)).

We will show that $p^* = p(1/2)$ is a threshold for $\mathcal{P}$. Let $G_1, G_2, \ldots, G_k$ be independent copies of $\mathbb{G}_{n,p}$. The graph $G_1 \cup G_2 \cup \cdots \cup G_k$ is distributed as $\mathbb{G}_{n,1-(1-p)^k}$. Now $1-(1-p)^k \leq kp$, and therefore by the coupling argument,

$$\mathbb{G}_{n,1-(1-p)^k} \subseteq \mathbb{G}_{n,kp},$$

and so $\mathbb{G}_{n,kp} \notin \mathcal{P}$ implies $G_1, G_2, \ldots, G_k \notin \mathcal{P}$. Hence

$$\mathbb{P}(\mathbb{G}_{n,kp} \notin \mathcal{P}) \leq [\mathbb{P}(\mathbb{G}_{n,p} \notin \mathcal{P})]^k.$$

Let ω be a function of n such that $\omega \to \infty$ arbitrarily slowly as $n \to \infty$, $\omega \ll \log\log n$.

Suppose also that $p = p^* = p(1/2)$ and $k = \omega$. Then

$$\mathbb{P}(\mathbb{G}_{n,\omega p^*} \notin \mathcal{P}) \leq 2^{-\omega} = o(1).$$

On the other hand, for $p = p^*/\omega$,

$$\frac{1}{2} = \mathbb{P}(\mathbb{G}_{n,p^*} \notin \mathcal{P}) \leq \left[\mathbb{P}(\mathbb{G}_{n,p^*/\omega} \notin \mathcal{P})\right]^{\omega}.$$

So

$$\mathbb{P}(\mathbb{G}_{n,p^*/\omega} \notin \mathcal{P}) \geq 2^{-1/\omega} = 1 - o(1).$$

□

In order to shorten many statements of theorems in the book, we say that a sequence of events $\mathcal{E}_n$ occurs *with high probability* (w.h.p.) if

$$\lim_{n\to\infty} \mathbb{P}(\mathcal{E}_n) = 1.$$

Thus the statement that says p^* is a threshold for a property $\mathcal{P}$ in $\mathbb{G}_{n,p}$ is the same as saying that $\mathbb{G}_{n,p} \notin \mathcal{P}$ w.h.p. if $p \ll p^*$, while $\mathbb{G}_{n,p} \in \mathcal{P}$ w.h.p. if $p \gg p^*$.

In the literature w.h.p. is often replaced by a.a.s. (*asymptotically almost surely*), not to be confused with a.s. (*almost surely*).

Exercises

3.2.1. Prove that if $\text{Var}\, X/\mathbb{E}\, X \to 0$ as $n \to \infty$, then for every $\varepsilon > 0$,

$$\mathbb{P}((1-\varepsilon)\,\mathbb{E}\, X < X < (1+\varepsilon)EX) \to 1.$$

3.2.2. Find in $\mathbb{G}_{n,p}$ and in $\mathbb{G}_{n,m}$ the expected number of maximal induced trees on $k \geq 2$ vertices. (Note that an induced tree is maximal in $\mathbb{G}_{n,p}$ if there is no vertex outside this tree connected to exactly one of its vertices.)

3.2.3. Suppose that $p = d/n$ where $d = o(n^{1/3})$. Show that w.h.p. $G_{n,p}$ has no copies of K_4.

3.2.4. Suppose that $p = d/n$ where d is a constant, $d > 1$. Show that w.h.p. $G_{n,p}$ contains an *induced* path of length $(\log n)^{1/2}$.

3.2.5. Suppose that $p = d/n$ where $d = O(1)$. Prove that w.h.p., in $G_{n,p}$, for all $S \subseteq [n], |S| \leq n/\log n$, we have $e(S) \leq 2|S|$, where $e(S)$ is the number of edges contained in S.

3.2.6. Suppose that $p = \log n/n$. Let a vertex of $G_{n,p}$ be small if its degree is less than $\log n/100$. Show that w.h.p. there is no edge of $G_{n,p}$ joining two small vertices.

3.2.7. Suppose that $p = d/n$ where d is constant. Prove that w.h.p., in $G_{n,p}$, no vertex belongs to more than one triangle.

3.2.8. Suppose that $p = d/n$ where d is constant. Prove that w.h.p. $G_{n,p}$ contains a vertex of degree exactly $\lceil (\log n)^{1/2} \rceil$.

3.2.9. Prove that if $np = \omega \leq \log n$, where $\omega = \omega(n) \to \infty$ as $n \to \infty$, then w.h.p. $\mathbb{G}_{n,p}$ contains at least one triangle. Use a coupling argument to show that it is also true for larger p.

3.2.10. Suppose that $k \geq 3$ is constant and that $np \to \infty$. Show that w.h.p. $\mathbb{G}_{n,p}$ contains a copy of the k-cycle, C_k.

3.2.11. Find the threshold for the existence in $\mathbb{G}_{n,p}$ of a copy of a diamond (a cycle on four vertices with a chord).

Problems for Chapter 3

3.1 Prove statement (i) of Theorem 1.4.

Let $0 \leq p_0 \leq 1$ and $N = \binom{n}{2}$. Suppose that $\mathcal{P}$ is a graph property such that $\mathbb{P}(\mathbb{G}_{n,m} \in \mathcal{P}) \to p_0$ for all $m = Np + O(\sqrt{Np(1-p)})$. Show that then $\mathbb{P}(\mathbb{G}_{n,p} \in \mathcal{P}) \to p_0$ as $n \to \infty$.

3.2 Prove statement (ii) of Theorem 1.4.

3.3 Let $\mathcal{P}$ be an increasing graph property, $N = \binom{n}{2}$ and let $0 \leq m \leq N$ while $p = m/N$. Assume that $\delta > 0$ is fixed and $0 \leq (1 \pm \delta)p \leq 1$. Show that

(1) if $\mathbb{P}(\mathbb{G}_{n,p} \in \mathcal{P}) \to 1$, then $\mathbb{P}(\mathbb{G}_{n,m} \in \mathcal{P}) \to 1$,

(2) if $\mathbb{P}(\mathbb{G}_{n,p} \in \mathcal{P}) \to 0$, then $\mathbb{P}(\mathbb{G}_{n,m} \in \mathcal{P}) \to 0$,

(3) if $\mathbb{P}(\mathbb{G}_{n,m} \in \mathcal{P}) \to 1$, then $\mathbb{P}(\mathbb{G}_{n,(1+\delta)p} \in \mathcal{P}) \to 1$,

(4) if $\mathbb{P}(\mathbb{G}_{n,m} \in \mathcal{P}) \to 0$, then $\mathbb{P}(\mathbb{G}_{n,(1-\delta)p} \in \mathcal{P}) \to 0$.

3.4 Let $\mathcal{P}$ be a monotone increasing property and let $m_1 < m_2$. Suppose that $\mathbb{P}(\mathbb{G}_{n,m_1} \in \mathcal{P}) < 1$ and $\mathbb{P}(\mathbb{G}_{n,m_2} \in \mathcal{P}) > 0$. Show that $\mathbb{P}(\mathbb{G}_{n,m_1} \in \mathcal{P}) < \mathbb{P}(\mathbb{G}_{n,m_2} \in \mathcal{P})$.

3.5 A graph property $\mathcal{P}$ is *convex* if graphs $G', G'' \in \mathcal{P}$ and $G' \subseteq G \subseteq G''$; then also $G \in \mathcal{P}$. Give at least two examples of convex graph properties and show that each such property $\mathcal{P}$ is an intersection of an increasing property $\mathcal{P}'$ and decreasing property $\mathcal{P}''$. Is it true in general?

3.6 Let $\mathcal{P}$ be a convex graph property graph and let m_1, m, m_2 be integer functions of n satisfying $0 \le m_1 \le m \le m_2 \le \binom{n}{2}$; then

$$\mathbb{P}(\mathbb{G}_{n,m} \in \mathcal{P}) \ge \mathbb{P}(\mathbb{G}_{n,m_1} \in \mathcal{P}) + \mathbb{P}(\mathbb{G}_{n,m_2} \in \mathcal{P}) - 1.$$

3.7 Let $\mathcal{P}$ be a convex graph property, $N = \binom{n}{2}$ and let $0 \le m \le N$ while $p = m/N$. Show that if $\mathbb{P}(\mathbb{G}_{n,p}) \to 1$ as $n \to \infty$, then $\mathbb{P}(\mathbb{G}_{n,m}) \to 1$.

4 Evolution

Here begins our story of the typical growth of a random graph. All the results up to Section 4.3 were first proved in a landmark paper by Erdős and Rényi [43]. The notion of the *evolution* of a random graph stems from a dynamic view of a *graph process*: viz. a sequence of graphs: $\mathbb{G}_0 = ([n], \emptyset), \mathbb{G}_1, \mathbb{G}_2, \ldots, \mathbb{G}_m, \ldots, \mathbb{G}_N = K_n$, where $\mathbb{G}_{m+1}$ is obtained from $\mathbb{G}_m$ by adding a random edge e_m. We see that there are $\binom{n}{2}!$ such sequences and $\mathbb{G}_m$ and $\mathbb{G}_{n,m}$ have the same distribution.

In the process of the evolution of a random graph we consider properties possessed by $\mathbb{G}_m$ or $\mathbb{G}_{n,m}$ w.h.p. when $m = m(n)$ grows from 0 to $\binom{n}{2}$, while in the case of $\mathbb{G}_{n,p}$ we analyze its typical structure when $p = p(n)$ grows from 0 to 1 as $n \to \infty$.

In the current chapter we mainly explore how the typical component structure evolves as the number of edges m increases. The following statements should be qualified with the caveat, w.h.p. The evolution of Erdős–Rényi-type random graphs has clearly distinguishable phases. The first phase, at the beginning of the evolution, can be described as a period when a random graph is a collection of small components which are mostly trees. Indeed the first result in this section shows that a random graph $\mathbb{G}_{n,m}$ is w.h.p. a collection of tree components as long as $m = o(n)$, or, equivalently, as long as $p = o(n^{-1})$ in $\mathbb{G}_{n,p}$. In more detail, we see that initially $G_m, m = o(n^{1/2})$ contains only isolated edges. Gradually larger and larger components appear, but while $m = o(n)$ we will see that G_m remains a forest. When $m = cn$ for some constant $c < 1/2$, cycles *may* appear but G_m consists of a forest with the maximum component size $O(\log n)$. There may be a few unicyclic components consisting of a tree plus an edge. No component contains more than one cycle. When $m \sim n/2$, things get very complicated, and when the process emerges with $m = cn, c > 1/2$, there is a unique giant component of size $\Omega(n)$ plus a forest with trees of maximum size $O(\log n)$ plus a few unicyclic components. This phase transition in the component structure is one of the most fascinating aspects of the process. We proceed to justify these statements.

4.1 Subcritical Phase

For clarity, all results presented in this chapter are stated in terms of $\mathbb{G}_{n,m}$. Due to the fact that computations are much easier for $\mathbb{G}_{n,p}$ we will first prove results in this model, and then the results for $\mathbb{G}_{n,m}$ will follow by the equivalence established either in

Lemmas 3.2 and 3.3 or in Theorem 3.4. We will also assume, throughout this chapter, that $\omega = \omega(n)$ is a function growing slowly with n, e.g., $\omega = \log\log n$ will suffice.

Theorem 4.1 *If $m \ll n$, then $\mathbb{G}_m$ is a forest w.h.p.*

Proof Suppose $m = n/\omega$ and let $N = \binom{n}{2}$, so $p = m/N \leq 3/(\omega n)$. Let X be the number of cycles in $\mathbb{G}_{n,p}$. Then

$$\mathbb{E}\,X = \sum_{k=3}^{n}\binom{n}{k}\frac{(k-1)!}{2}p^k \leq \sum_{k=3}^{n}\frac{n^k}{k!}\frac{(k-1)!}{2}p^k \leq \sum_{k=3}^{n}\frac{n^k}{2k}\frac{3^k}{\omega^k n^k} = O(\omega^{-3}) \to 0.$$

Therefore, by the First Moment Method (see (3.11)),

$$\mathbb{P}(\mathbb{G}_{n,p} \text{ is not a forest}) = \mathbb{P}(X \geq 1) \leq \mathbb{E}\,X = o(1),$$

which implies that

$$\mathbb{P}(\mathbb{G}_{n,p} \text{ is a forest}) \to 1 \text{ as } n \to \infty.$$

Notice that the property that a graph is a forest is monotone decreasing, so by Lemma 3.3,

$$\mathbb{P}(\mathbb{G}_m \text{ is a forest}) \to 1 \text{ as } n \to \infty.$$

(Note that we have actually used Lemma 3.3 to show that $\mathbb{P}(\mathbb{G}_{n,p}$ is not a forest) $=o(1)$, which implies that $\mathbb{P}(G_m$ is not a forest)$=o(1)$.) □

As we keep adding edges, trees on more and more vertices gradually start to appear. The next two theorems show how long we have to "wait" until trees with a given number of vertices appear w.h.p.

Theorem 4.2 *Fix $k \geq 3$. If $m \ll n^{\frac{k-2}{k-1}}$, then w.h.p. $\mathbb{G}_m$ contains no tree with k vertices.*

Proof Let $m = n^{\frac{k-2}{k-1}}/\omega$ and then

$$p = \frac{m}{N} \sim \frac{2}{\omega n^{k/(k-1)}} \leq \frac{3}{\omega n^{k/(k-1)}}.$$

Let X_k denote the number of trees with k vertices in $\mathbb{G}_{n,p}$. Let $T_1, T_2, \ldots, T_M$ be an enumeration of the copies of k-vertex trees in K_n. Let

$$A_i = \{T_i \text{ occurs as a subgraph in } \mathbb{G}_{n,p}\}.$$

The probability that a tree T occurs in $\mathbb{G}_{n,p}$ is $p^{e(T)}$, where $e(T)$ is the number of edges of T. So,

$$\mathbb{E}\,X_k = \sum_{t=1}^{M}\mathbb{P}(A_t) = Mp^{k-1}.$$

But $M = \binom{n}{k}k^{k-2}$ since one can choose a set of k vertices in $\binom{n}{k}$ ways and then by Cayley's formula choose a tree on these vertices in k^{k-2} ways. Hence

$$\mathbb{E}\, X_k = \binom{n}{k} k^{k-2} p^{k-1}. \tag{4.1}$$

Noting also that, by (2.2), for every n and k, $\binom{n}{k} \leq \left(\frac{ne}{k}\right)^k$, we see that

$$\begin{aligned}\mathbb{E}\, X_k &\leq \left(\frac{ne}{k}\right)^k k^{k-2} \left(\frac{3}{\omega n^{k/(k-1)}}\right)^{k-1} \\ &= \frac{3^{k-1} e^k}{k^2 \omega^{k-1}} \to 0,\end{aligned}$$

as $n \to \infty$, seeing as k is fixed.

Thus we see by the First Moment Method that

$$\mathbb{P}(\mathbb{G}_{n,p} \text{ contains a tree with } k \text{ vertices}) \to 0.$$

This property is monotone increasing and therefore

$$\mathbb{P}(\mathbb{G}_m \text{ contains a tree with } k \text{ vertices}) \to 0. \qquad \square$$

Let us check what happens if the number of edges in $\mathbb{G}_m$ is much larger than $n^{\frac{k-2}{k-1}}$.

Theorem 4.3 *Fix $k \geq 3$. If $m \gg n^{\frac{k-2}{k-1}}$, then w.h.p. $\mathbb{G}_m$ contains a copy of every fixed tree with k vertices.*

Proof Let $p = \frac{m}{N}, m = \omega n^{\frac{k-2}{k-1}}$, where $\omega = o(\log n)$ and fix some tree T with k vertices. Denote by $\hat{X}_k$ the number of *isolated* copies of T (T-components) in $\mathbb{G}_{n,p}$. Let $\operatorname{aut}(H)$ denote the number of automorphisms of a graph H. Note that there are $k!/\operatorname{aut}(T)$ copies of T in the complete graph K_k. To see this choose a copy of T with vertex set $[k]$. There are $k!$ ways of mapping the vertices of T to the vertices of K_k. Each map f induces a copy of T and two maps f_1, f_2 induce the same copy if and only if $f_2 f_1^{-1}$ is an automorphism of T.

So,

$$\mathbb{E}\, \hat{X}_k = \binom{n}{k} \frac{k!}{\operatorname{aut}(T)} p^{k-1} (1-p)^{k(n-k)+\binom{k}{2}-k+1} \tag{4.2}$$

$$= (1+o(1)) \frac{(2\omega)^{k-1}}{\operatorname{aut}(T)} \to \infty. \tag{4.3}$$

In (4.2) we have approximated $\binom{n}{k} \leq \frac{n^k}{k!}$ and used the fact that $\omega = o(\log n)$ in order to show that $(1-p)^{k(n-k)+\binom{k}{2}-k+1} = 1 - o(1)$.

Next let $\mathcal{T}$ be the set of copies of T in K_n and $T_{[k]}$ be a fixed copy of T on vertices $[k]$ of K_n. Then,

$$\mathbb{E}(\hat{X}_k^2) = \sum_{T_1, T_2 \in \mathcal{T}} \mathbb{P}(T_2 \subseteq_i \mathbb{G}_{n,p} | \, T_1 \subseteq_i \mathbb{G}_{n,p}) \, \mathbb{P}(T_1 \subseteq_i \mathbb{G}_{n,p})$$

$$= \mathbb{E}\,\hat{X}_k \left(1 + \sum_{\substack{T_2 \in \mathcal{T} \\ V(T_2) \cap [k] = \emptyset}} \mathbb{P}(T_2 \subseteq_i \mathbb{G}_{n,p} | \, T_{[k]} \subseteq_i \mathbb{G}_{n,p}) \right)$$

$$\leq \mathbb{E}\,\hat{X}_k \left(1 + (1-p)^{-k^2} \,\mathbb{E}\, X_k \right).$$

Notice that the $(1-p)^{-k^2}$ factor comes from conditioning on the event $T_{[k]} \subseteq_i \mathbb{G}_{n,p}$, which forces the nonexistence of fewer than k^2 edges.

Hence, by the Second Moment Method,

$$\mathbb{P}(\hat{X}_k > 0) \geq \frac{(\mathbb{E}\,\hat{X}_k)^2}{\mathbb{E}\,\hat{X}_k \left(1 + (1-p)^{-k^2}\,\mathbb{E}\,\hat{X}_k\right)} \to 1$$

as $n \to \infty$, since $p \to 0$ and $\mathbb{E}\,\hat{X}_k \to \infty$.

Thus

$$\mathbb{P}(\mathbb{G}_{n,p} \text{ contains a copy of isolated tree } T) \to 1,$$

which implies that

$$\mathbb{P}(\mathbb{G}_{n,p} \text{ contains a copy of } T) \to 1.$$

As the property of having a copy of a tree T is monotone increasing, it in turn implies that

$$\mathbb{P}(\mathbb{G}_m \text{ contains a copy of } T) \to 1$$

as $m \gg n^{\frac{k-2}{k-1}}$ and $n \to \infty$. □

Combining the above two theorems we arrive at the following conclusion.

Corollary 4.4 *The function $m^*(n) = n^{\frac{k-2}{k-1}}$ is the threshold for the property that a random graph $\mathbb{G}_m$ contains a tree with $k \geq 3$ vertices, i.e.,*

$$\mathbb{P}(\mathbb{G}_m \supseteq k\text{-vertex tree}) = \begin{cases} o(1) & \text{if } m \ll n^{\frac{k-2}{k-1}}, \\ 1 - o(1) & \text{if } m \gg n^{\frac{k-2}{k-1}}. \end{cases}$$

We complete our presentation of the basic features of a random graph in its subcritical phase of evolution with a description of the order of its largest component.

Theorem 4.5 *If $m = \frac{1}{2}cn$, where $0 < c < 1$ is a constant, then w.h.p. the order of the largest component of a random graph $\mathbb{G}_m$ is $O(\log n)$.*

The above theorem follows from the next three lemmas stated and proved in terms of $\mathbb{G}_{n,p}$ with $p = c/n$, $0 < c < 1$. In fact the first of those three lemmas covers a little bit more than the case of $p = c/n$, $0 < c < 1$.

Lemma 4.6 *If* $p \le \frac{1}{n} - \frac{\omega}{n^{4/3}}$*, where* $\omega = \omega(n) \to \infty$*, then w.h.p. every component in* $\mathbb{G}_{n,p}$ *contains at most one cycle.*

Proof Suppose that there is a pair of cycles that are in the same component. If such a pair exists, then there is *minimal* pair C_1, C_2, i.e., either C_1 and C_2 are connected by a path (or meet at a vertex) or they form a cycle with a diagonal path (see Figure 4.1). Then in either case, $C_1 \cup C_2$ consists of a path P plus another two distinct edges, one from each endpoint of P joining it to another vertex in P. The number of such graphs on k labeled vertices can be bounded by $k^2 k!$.

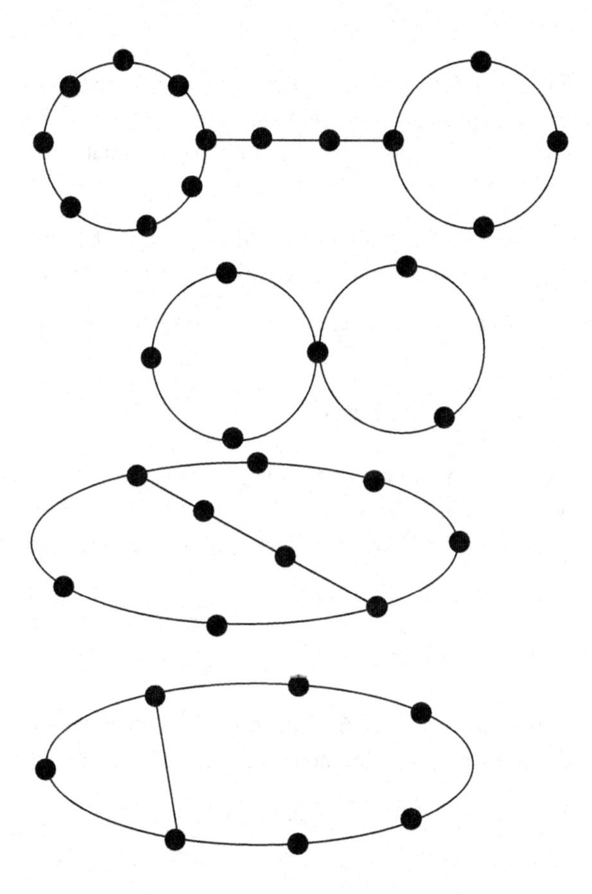

Figure 4.1 $C_1 \cup C_2$

Let X be the number of subgraphs of the above kind (shown in Figure 4.1) in the random graph $\mathbb{G}_{n,p}$. By the First Moment Method,

$$\mathbb{P}(X > 0) \le \mathbb{E}\,X \le \sum_{k=4}^{n} \binom{n}{k} k^2 k! p^{k+1} \tag{4.4}$$

$$\leq \sum_{k=4}^{n} \frac{n^k}{k!} k^2 k! \frac{1}{n^{k+1}} \left(1 - \frac{\omega}{n^{1/3}}\right)^{k+1}$$
$$\leq \int_0^\infty \frac{x^2}{n} \exp\left\{-\frac{\omega x}{n^{1/3}}\right\} dx = \frac{2}{\omega^3} = o(1).$$

□

We remark for later use that if $p = c/n$, $0 < c < 1$, then (4.4) implies that

$$\mathbb{P}(X > 0) \leq \sum_{k=4}^{n} k^2 c^{k+1} n^{-1} = O(n^{-1}). \tag{4.5}$$

Hence, in determining the order of the largest component we may concentrate our attention on unicyclic components and tree components (isolated trees). However, the number of vertices on unicyclic components tends to be rather small, as is shown in the next lemma.

Lemma 4.7 *If $p = c/n$, where $c \neq 1$ is a constant, then in $\mathbb{G}_{n,p}$ w.h.p. the number of vertices in components with exactly one cycle is $O(\omega)$ for any growing function ω.*

Proof Let X_k be the number of vertices on unicyclic components with k vertices. Then

$$\mathbb{E} X_k \leq \binom{n}{k} k^{k-2} \binom{k}{2} k p^k (1-p)^{k(n-k)+\binom{k}{2}-k}. \tag{4.6}$$

The factor $k^{k-2}\binom{k}{2}$ in (4.6) is the number of choices for a tree plus an edge on k vertices in $[k]$. This bounds the number $C(k,k)$ of connected graphs on $[k]$ with k edges. This is off by a factor $O(k^{1/2})$ from the exact formula which is given below for completeness:

$$C(k,k) = \sum_{r=3}^{k} \binom{k}{r} \frac{(r-1)!}{2} r k^{k-r-1} \sim \sqrt{\frac{\pi}{8}}\, k^{k-1/2}. \tag{4.7}$$

The remaining factor, other than $\binom{n}{k}$, in (4.6) is the probability that the k edges of the unicyclic component exist and that there are no other edges on $\mathbb{G}_{n,p}$ incident with the k chosen vertices.

Note also that, by (2.4), for every n and k, $\binom{n}{k} \leq \frac{n^k}{k!} e^{-k(k-1)/2n}$. Assume next that $c < 1$ and then we get

$$\mathbb{E} X_k \leq \frac{n^k}{k!} e^{-\frac{k(k-1)}{2n}} k^{k+1} \frac{c^k}{n^k} e^{-ck+\frac{ck(k-1)}{2n}+\frac{ck}{2n}} \tag{4.8}$$
$$\leq \frac{e^k}{k^k} e^{-\frac{k(k-1)}{2n}} k^{k+1} c^k e^{-ck+\frac{k(k-1)}{2n}+\frac{c}{2}} \tag{4.9}$$
$$\leq k\left(ce^{1-c}\right)^k e^{\frac{c}{2}}.$$

So,

$$\mathbb{E} \sum_{k=3}^{n} X_k \leq \sum_{k=3}^{n} k\left(ce^{1-c}\right)^k e^{\frac{c}{2}} = O(1), \tag{4.10}$$

since $ce^{1-c} < 1$ for $c \neq 1$. By the Markov inequality, if $\omega = \omega(n) \to \infty$ (see Lemma 2.15),

$$\mathbb{P}\left(\sum_{k=3}^{n} X_k \geq \omega\right) = O\left(\frac{1}{\omega}\right) \to 0 \text{ as } n \to \infty,$$

and the lemma follows for $c < 1$.

If $c > 1$, then we cannot deduce (4.9) from (4.8). If however $k = o(n)$, then this does not matter, since then $e^{k^2/n} = e^{o(k)}$. In the proof of Theorem 4.10 below we show that when $c > 1$, there is w.h.p. a unique giant component of size $\Omega(n)$ and all other components are of size $O(\log n)$. This giant is not unicyclic. This enables us to complete the proof of this lemma for $c > 1$. □

After proving the first two lemmas one can easily see that the only remaining candidate for the largest component of our random graph is an isolated tree.

Lemma 4.8 *Let $p = \frac{c}{n}$, where $c \neq 1$ is a constant, $\alpha = c - 1 - \log c$, and $\omega = \omega(n) \to \infty$, $\omega = o(\log\log n)$. Then*

(i) w.h.p. there exists an isolated tree of order

$$k_- = \frac{1}{\alpha}\left(\log n - \frac{5}{2}\log\log n\right) - \omega,$$

(ii) w.h.p. there is no *isolated tree of order at least*

$$k_+ = \frac{1}{\alpha}\left(\log n - \frac{5}{2}\log\log n\right) + \omega.$$

Proof Note that our assumption on c means that α is a positive constant.

Let X_k be the number of isolated trees of order k. Then

$$\mathbb{E}\, X_k = \binom{n}{k} k^{k-2} p^{k-1} (1-p)^{k(n-k)+\binom{k}{2}-k+1}. \tag{4.11}$$

To prove (i) suppose $k = O(\log n)$. Then $\binom{n}{k} \sim \frac{n^k}{k!}$ and by using inequalities (3.9), (3.10) and Stirling's approximation, Lemma 2.4, for $k!$, we see that

$$\begin{aligned}
\mathbb{E}\, X_k &= (1+o(1))\frac{n}{c}\frac{k^{k-2}}{k!}(ce^{-c})^k \qquad (4.12)\\
&= \frac{(1+o(1))}{c\sqrt{2\pi}}\frac{n}{k^{5/2}}(ce^{1-c})^k\\
&= \frac{(1+o(1))}{c\sqrt{2\pi}}\frac{n}{k^{5/2}}e^{-\alpha k} \qquad \text{for } k = O(\log n). \qquad (4.13)
\end{aligned}$$

Putting $k = k_-$ we see that

$$\mathbb{E}\, X_k = \frac{(1+o(1))}{c\sqrt{2\pi}}\frac{n}{k^{5/2}}\frac{e^{\alpha\omega}(\log n)^{5/2}}{n} \geq Ae^{\alpha\omega} \tag{4.14}$$

for some constant $A > 0$.

We continue via the Second Moment Method, this time using the Chebyshev inequality as we will need a little extra precision for the proof of Theorem 4.10. Using essentially the same argument as for a fixed tree T of order k (see Theorem 4.3), we get

$$\mathbb{E}\, X_k^2 \le \mathbb{E}\, X_k \left(1 + (1-p)^{-k^2}\, \mathbb{E}\, X_k\right).$$

So

$$\begin{aligned} \operatorname{Var} X_k &\le \mathbb{E}\, X_k + (\mathbb{E}\, X_k)^2 \left((1-p)^{-k^2} - 1\right) \\ &\le \mathbb{E}\, X_k + 2ck^2 (\mathbb{E}\, X_k)^2/n \qquad \text{for } k = O(\log n). \end{aligned} \tag{4.15}$$

Thus, by the Chebyshev inequality (see Lemma 2.18), we see that for any $\varepsilon > 0$,

$$\mathbb{P}\left(|X_k - \mathbb{E}\, X_k| \ge \varepsilon\, \mathbb{E}\, X_k\right) \le \frac{1}{\varepsilon^2\, \mathbb{E}\, X_k} + \frac{2ck^2}{\varepsilon^2 n} = o(1). \tag{4.16}$$

Thus w.h.p. $X_k \ge A e^{\alpha\omega/2}$ and this completes the proof of (i).

For (ii) we go back to formula (4.11) and write, for some new constant $A > 0$,

$$\begin{aligned} \mathbb{E}\, X_k &\le \frac{A}{\sqrt{k}} \left(\frac{ne}{k}\right)^k k^{k-2} \left(1 - \frac{k}{2n}\right)^{k-1} \left(\frac{c}{n}\right)^{k-1} e^{-ck + \frac{ck^2}{2n}} \\ &\le \frac{2An}{\widehat{c}_k k^{5/2}} \left(\widehat{c}_k e^{1-\widehat{c}_k}\right)^k, \end{aligned}$$

where $\widehat{c}_k = c\left(1 - \frac{k}{2n}\right)$.

In the case $c < 1$ we have $\widehat{c}_k e^{1-\widehat{c}_k} \le c e^{1-c}$ and $\widehat{c}_k \sim c$, and so we can write

$$\begin{aligned} \sum_{k=k_+}^{n} \mathbb{E}\, X_k &\le \frac{3An}{c} \sum_{k=k_+}^{n} \frac{\left(ce^{1-c}\right)^k}{k^{5/2}} \le \frac{3An}{ck_+^{5/2}} \sum_{k=k_+}^{\infty} e^{-\alpha k} \\ &= \frac{3Ane^{-\alpha k_+}}{ck_+^{5/2}(1 - e^{-\alpha})} = \frac{(3A + o(1))\alpha^{5/2} e^{-\alpha\omega}}{c(1 - e^{-\alpha})} = o(1). \end{aligned} \tag{4.17}$$

If $c > 1$, then for $k \le \frac{n}{\log n}$ we use $\widehat{c}_k e^{1-\widehat{c}_k} = e^{-\alpha - O(1/\log n)}$ and for $k > \frac{n}{\log n}$ we use $c_k \ge c/2$ and $\widehat{c}_k e^{1-\widehat{c}_k} \le 1$ and replace (4.17) by

$$\sum_{k=k_+}^{n} \mathbb{E}\, X_k \le \frac{3An}{ck_+^{5/2}} \sum_{k=k_+}^{n/\log n} e^{-(\alpha + O(1/\log n))k} + \frac{6An}{c} \sum_{k=n/\log n}^{n} \frac{1}{k^{5/2}} = o(1).$$

□

Finally, applying Lemmas 4.7 and 4.8 we can prove the following useful identity: Suppose that $x = x(c)$ is given as

$$x = x(c) = \begin{cases} c & c \le 1, \\ \text{the solution in } (0,1) \text{ to } xe^{-x} = ce^{-c} & c > 1. \end{cases}$$

Note that xe^{-x} increases continuously as x increases from 0 to 1 and then decreases. This justifies the existence and uniqueness of x.

Lemma 4.9 *If $c > 0,\ c \neq 1$ is a constant, and $x = x(c)$ is as defined above, then*

$$\frac{1}{x}\sum_{k=1}^{\infty}\frac{k^{k-1}}{k!}\left(ce^{-c}\right)^k = 1.$$

Proof Let $p = \frac{c}{n}$. Assume first that $c < 1$ and let X be the total number of vertices of $\mathbb{G}_{n,p}$ that lie in nontree components. Let X_k be the number of tree components of order k. Then,

$$n = \sum_{k=1}^{n} kX_k + X.$$

So,

$$n = \sum_{k=1}^{n} k\,\mathbb{E}\,X_k + \mathbb{E}\,X.$$

Now,

(i) by (4.5) and (4.10), $\mathbb{E}\,X = O(1)$,
(ii) by (4.12), if $k < k_+$, then

$$\mathbb{E}\,X_k = (1+o(1))\frac{n}{ck!}k^{k-2}\left(ce^{-c}\right)^k.$$

So, by Lemma 4.8,

$$\begin{aligned} n &= o(n) + \frac{n}{c}\sum_{k=1}^{k_+}\frac{k^{k-1}}{k!}\left(ce^{-c}\right)^k \\ &= o(n) + \frac{n}{c}\sum_{k=1}^{\infty}\frac{k^{k-1}}{k!}\left(ce^{-c}\right)^k. \end{aligned}$$

Now divide through by n and let $n \to \infty$.

This proves the identity for the case $c < 1$. Suppose now that $c > 1$. Then, since x is a solution of the equation $ce^{-c} = xe^{-x}$, $0 < x < 1$, we have

$$\sum_{k=1}^{\infty}\frac{k^{k-1}}{k!}\left(ce^{-c}\right)^k = \sum_{k=1}^{\infty}\frac{k^{k-1}}{k!}\left(xe^{-x}\right)^k = x$$

by the first part of the proof (for $c < 1$). □

We note that, in fact, Lemma 4.9 is also true for $c = 1$.

Exercises

4.1.1. Verify equation (4.3).

4.1.2. Find an upper bound on the expected number of cycles on k vertices in $\mathbb{G}_{n,p}$ and show its asymptotic behavior ($n \to \infty$) when $p = c/n$ and $c < 1, c = 1$ and $c > 1$.

4.1.3. Prove that if $m \ll n^{1/2}$, then $w.h.p.$ the random graph $\mathbb{G}_{n,m}$ is the union of isolated vertices and edges only.

4.1.4. Prove that if $m \gg n^{1/2}$, then $\mathbb{G}_{n,m}$ contains a path of length 2 w.h.p.

4.1.5. Show that if $m = cn^{(k-2)(k-1)}$, where $c > 0$, and T is a fixed tree with $k \geq 3$ vertices, then

$$\mathbb{P}(\mathbb{G}_{n,m} \text{ contains an isolated copy of tree } T) \to 1 - e^{-\lambda}$$

as $n \to \infty$, where $\lambda = (2c)^{k-1}/\mathrm{aut}(T)$ and $\mathrm{aut}(T)$ denotes the number of automorphisms of T.

4.1.6. Show that if $c \neq 1$ and $xe^{-x} = ce^{-c}$, where $0 < x < 1$, then

$$\frac{1}{c}\sum_{k=1}^{\infty} \frac{k^{k-2}}{k!}(ce^{-c})^k = \begin{cases} 1 - \frac{c}{2} & c < 1, \\ \frac{x}{c}\left(1 - \frac{x}{2}\right) & c > 1. \end{cases}$$

4.2 Supercritical Phase

The structure of a random graph $\mathbb{G}_m$ changes dramatically when $m = \frac{1}{2}cn$, where $c > 1$ is a constant. We will give a precise characterization of this phenomenon, presenting results in terms of $\mathbb{G}_m$ and proving them for $\mathbb{G}_{n,p}$ with $p = c/n,\ c > 1$.

Theorem 4.10 *If $m = cn/2, c > 1$, then w.h.p. $\mathbb{G}_m$ consists of a unique giant component, with $\left(1 - \frac{x}{c} + o(1)\right)n$ vertices and $\left(1 - \frac{x^2}{c^2} + o(1)\right)\frac{cn}{2}$ edges. Here $0 < x < 1$ is the solution of the equation $xe^{-x} = ce^{-c}$. The remaining components are of order at most $O(\log n)$.*

Proof Suppose that Z_k is the number of components of order k in $\mathbb{G}_{n,p}$. Then, bounding the number of such components by the number of trees with k vertices that span a component, we get

$$\begin{aligned} \mathbb{E}\, Z_k &\leq \binom{n}{k} k^{k-2} p^{k-1} (1-p)^{k(n-k)} \qquad (4.18) \\ &\leq A\left(\frac{ne}{k}\right)^k k^{k-2}\left(\frac{c}{n}\right)^{k-1} e^{-ck+ck^2/n} \\ &\leq \frac{An}{k^2}\left(ce^{1-c+ck/n}\right)^k. \end{aligned}$$

Now let $\beta_1 = \beta_1(c)$ be small enough so that

$$ce^{1-c+c\beta_1} < 1,$$

and let $\beta_0 = \beta_0(c)$ be large enough so that

$$\left(ce^{1-c+o(1)}\right)^{\beta_0 \log n} < \frac{1}{n^2}.$$

If we choose β_1 and β_0 as above, then it follows that w.h.p. there is no component of order $k \in [\beta_0 \log n, \beta_1 n]$.

Our next task is to estimate the number of vertices on small components, i.e., those of size at most $\beta_0 \log n$.

We first estimate the total number of vertices on small tree components, i.e., on isolated trees of order at most $\beta_0 \log n$.

Assume first that $1 \le k \le k_0$, where $k_0 = \frac{1}{2\alpha} \log n$, where α is from Lemma 4.8. It follows from (4.12) that

$$\mathbb{E}\left(\sum_{k=1}^{k_0} kX_k\right) \sim \frac{n}{c} \sum_{k=1}^{k_0} \frac{k^{k-1}}{k!} (ce^{-c})^k$$
$$\sim \frac{n}{c} \sum_{k=1}^{\infty} \frac{k^{k-1}}{k!} (ce^{-c})^k$$

using $k^{k-1}/k! < e^k$, and $ce^{-c} < e^{-1}$ for $c \neq 1$ to extend the summation from k_0 to infinity.

Putting $\varepsilon = 1/\log n$ and using (4.16) we see that the probability that any X_k, $1 \le k \le k_0$, deviates from its mean by more than $1 \pm \varepsilon$ is at most

$$\sum_{k=1}^{k_0} \left[\frac{(\log n)^2}{n^{1/2-o(1)}} + O\left(\frac{(\log n)^4}{n}\right)\right] = o(1),$$

where the $n^{1/2-o(1)}$ term comes from putting $\omega \sim k_0/2$ in (4.14), which is allowed by (4.13) and (4.15).

Thus, if $x = x(c)$, $0 < x < 1$ is the unique solution in $(0,1)$ of the equation $xe^{-x} = ce^{-c}$, then w.h.p.

$$\sum_{k=1}^{k_0} kX_k \sim \frac{n}{c} \sum_{k=1}^{\infty} \frac{k^{k-1}}{k!} (xe^{-x})^k$$
$$= \frac{nx}{c}$$

by Lemma 4.9.

Now consider $k_0 < k \le \beta_0 \log n$. It follows from (4.11) that

$$\mathbb{E}\left(\sum_{k=k_0+1}^{\beta_0 \log n} kX_k\right) \le \sum_{k=k_0+1}^{\beta_0 \log n} \left(\frac{ne}{k}\right)^k k^{k-2} \left(\frac{c}{n}\right)^{k-1} e^{-ck(n-k)/n}$$
$$= O\left(n^{1/2+o(1)}\right). \tag{4.19}$$

So, by the Markov inequality (see Lemma 2.15), w.h.p.,

$$\sum_{k=k_0+1}^{\beta_0 \log n} kX_k = o(n).$$

Now consider the number Y_k of nontree components with k vertices, $1 \le k \le \beta_0 \log n$.

$$\mathbb{E}\left(\sum_{k=1}^{\beta_0 \log n} kY_k\right) \le \sum_{k=1}^{\beta_0 \log n} \binom{n}{k} k^{k-1} \binom{k}{2} \left(\frac{c}{n}\right)^k \left(1 - \frac{c}{n}\right)^{k(n-k)}$$

$$= O(1). \tag{4.20}$$

So, again by the Markov inequality, w.h.p.,

$$\sum_{k=1}^{\beta_0 \log n} kY_k = o(n).$$

Summarizing, we have proved so far that w.h.p. there are approximately $\frac{nx}{c}$ vertices on components of order k, where $1 \leq k \leq \beta_0 \log n$ and all the remaining *giant* components are of size at least $\beta_1 n$.

We complete the proof by showing the uniqueness of the giant component. Let

$$c_1 = c - \frac{\log n}{n} \text{ and } p_1 = \frac{c_1}{n}.$$

Define p_2 by

$$1 - p = (1 - p_1)(1 - p_2)$$

and note that $p_2 \geq \frac{\log n}{n^2}$. Then, see Section 3.2,

$$\mathbb{G}_{n,p} = \mathbb{G}_{n,p_1} \cup \mathbb{G}_{n,p_2}.$$

If $x_1 e^{-x_1} = c_1 e^{-c_1}$, then $x_1 \sim x$ and so, by our previous analysis, w.h.p., $\mathbb{G}_{n,p_1}$ has no components with the number of vertices in the range $[\beta_0 \log n, \beta_1 n]$.

Suppose there are components $C_1, C_2, \ldots, C_l$ with $|C_i| > \beta_1 n$. Here $l \leq 1/\beta_1$. Now we add edges of $\mathbb{G}_{n,p_2}$ to $\mathbb{G}_{n,p_1}$. Then

$$\begin{aligned}
\mathbb{P}\left(\exists i, j: \text{ no } \mathbb{G}_{n,p_2} \text{ edge joins } C_i \text{ with } C_j\right) &\leq \binom{l}{2}(1 - p_2)^{(\beta_1 n)^2} \\
&\leq l^2 e^{-\beta_1^2 \log n} \\
&= o(1).
\end{aligned}$$

So w.h.p. $\mathbb{G}_{n,p}$ has a unique component with more than $\beta_0 \log n$ vertices and it has $\sim \left(1 - \frac{x}{c}\right) n$ vertices.

We now consider the number of edges in the giant C_0. Now we switch to $G = \mathbb{G}_{n,m}$. Suppose that the edges of G are $e_1, e_2, \ldots, e_m$ in random order. We estimate the probability that $e = e_m = \{x, y\}$ is an edge of the giant. Let G_1 be the graph induced by $\{e_1, e_2, \ldots, e_{m-1}\}$. G_1 is distributed as $\mathbb{G}_{n,m-1}$, and so we know that w.h.p. G_1 has a unique giant C_1 and other components are of size $O(\log n)$. So the probability that e is an edge of the giant is $o(1)$ plus the probability that x or y is a vertex of C_1. Thus,

$$\begin{aligned}
\mathbb{P}\left(e \notin C_0 \mid |C_1| \sim n\left(1 - \frac{x}{c}\right)\right) &= \mathbb{P}\left(e \cap C_1 = \emptyset \mid |C_1| \sim n\left(1 - \frac{x}{c}\right)\right) \\
&= \left(1 - \frac{|C_1|}{n}\right)\left(1 - \frac{|C_1| - 1}{n}\right) \sim \left(\frac{x}{c}\right)^2.
\end{aligned} \tag{4.21}$$

It follows that the expected number of edges in the giant is as claimed. To prove concentration, it is simplest to use the Chebyshev inequality, see Lemma 2.18. So, now fix $i, j \leq m$ and let C_2 denote the unique giant component of $G_{n,m} - \{e_i, e_j\}$. Then, arguing as for (4.21),

$$\mathbb{P}(e_i, e_j \subseteq C_0) = o(1) + \mathbb{P}(e_j \cap C_2 \neq \emptyset \mid e_i \cap C_2 \neq \emptyset)\,\mathbb{P}(e_i \cap C_2 \neq \emptyset)$$
$$= (1 + o(1))\,\mathbb{P}(e_i \subseteq C_0)\,\mathbb{P}(e_j \subseteq C_0).$$

In the $o(1)$ term, we hide the probability of the event

$$\{e_i \cap C_2 \neq \emptyset, e_j \cap C_2 \neq \emptyset, e_i \cap e_j \neq \emptyset\},$$

which has probability $o(1)$. We should double this $o(1)$ probability here to account for switching the roles of i, j.

The Chebyshev inequality can now be used to show that the number of edges is concentrated as claimed. □

From the above theorem and the results of previous sections, we see that, when $m = cn/2$ and c passes the critical value equal to 1, the typical structure of a random graph changes from a scattered collection of small trees and unicyclic components to a coagulated lump of components (the giant component) that dominates the graph. This short period when the giant component emerges is called the *phase transition*. We will look at this fascinating period of the evolution more closely in Section 4.3.

We know that w.h.p. the giant component of $\mathbb{G}_{n,m}$, $m = cn/2$, $c > 1$ has $\sim \left(1 - \frac{x}{c}\right) n$ vertices and $\sim \left(1 - \frac{x^2}{c^2}\right) \frac{cn}{2}$ edges. So, if we look at the graph H induced by the vertices outside the giant, then w.h.p. H has $\sim n_1 = \frac{nx}{c}$ vertices and $\sim m_1 = xn_1/2$ edges. Thus we should expect H to resemble $\mathbb{G}_{n_1.m_1}$, which is subcritical since $x < 1$. This can be made precise, but the intuition is clear.

Now increase m further and look on the outside of the giant component. The giant component subsequently consumes the small components not yet attached to it. When m is such that $m/n \to \infty$, then unicyclic components disappear and a random graph $\mathbb{G}_m$ achieves the structure described in the next theorem.

Theorem 4.11 *Let $\omega = \omega(n) \to \infty$ as $n \to \infty$ be some slowly growing function. If $m \geq \omega n$ but $m \leq n(\log n - \omega)/2$, then $\mathbb{G}_m$ is disconnected and all components, with the exception of the giant, are trees w.h.p.*

Tree components of order k die out in the reverse order they were born, i.e., larger trees are "swallowed" by the giant earlier than smaller ones.

Exercises

4.2.1. Show that if $p = \omega/n$ where $\omega = \omega(n) \to \infty$, then w.h.p. $\mathbb{G}_{n,p}$ contains no unicyclic components. (A component is *unicyclic* if it contains exactly one cycle, i.e., is a tree plus one extra edge.)

4.2.2. Suppose that $np \to \infty$ and $3 \le k = O(1)$. Show that $\mathbb{G}_{n,p}$ contains a k-cycle w.h.p.

4.2.3. Verify equation (4.19).

4.2.4. Verify equation (4.20).

4.3 Phase Transition

In the previous two sections we studied the asymptotic behavior of $\mathbb{G}_m$ (and $\mathbb{G}_{n,p}$) in the "subcritical phase" when $m = cn/2, c < 1$ $(p = c/n, c < 1)$, as well as in the "supercritical phase" when $m = cn/2, c > 1$ $(p = c/n, c > 1)$ of its evolution.

We have learned that when $m = cn/2, c < 1$, our random graph consists w.h.p. of tree components and components with exactly one cycle (see Theorem 4.1 and Lemma 4.7). We call such components *simple*, while components which are not simple, i.e., components with at least two cycles, will be called *complex.*

All components during the subcritical phase are rather small, of order $\log n$, tree components dominate the typical structure of $\mathbb{G}_m$, and there is no significant gap in the order of the first and the second largest component. This follows from Lemma 4.8. The proof of this lemma shows that w.h.p. there are many trees of height k_-. The situation changes when $m > n/2$, i.e., when we enter the supercritical phase and then w.h.p. $\mathbb{G}_m$ consists of a single giant complex component (of the order comparable to n), and some number of simple components, i.e., tree components and components with exactly one cycle (see Theorem 4.10). One can also observe a clear gap between the order of the largest component (the giant) and the second largest component which is of the order $O(\log n)$. This phenomenon of dramatic change of the typical structure of a random graph is called its *phase transition.*

A natural question arises as to what happens when $m/n \to 1/2$, either from below or above, as $n \to \infty$. It appears that one can establish a so-called *scaling window* or *critical window* for the phase transition in which $\mathbb{G}_m$ is undergoing a rapid change in its typical structure. A characteristic feature of this period is that a random graph can w.h.p. consist of more than one complex component (recall that there are no complex components in the subcritical phase and there is a unique complex component in the supercritical phase).

Erdős and Rényi [43] studied the size of the largest tree in the random graph $\mathbb{G}_{n,m}$ when $m = n/2$ and showed that it was likely to be around $n^{2/3}$. They called the transition from $O(\log n)$ through $\Theta(n^{2/3})$ to $\Omega(n)$ the "double jump." They did not study the regime $m = n/2 + o(n)$. Bollobás [20] opened the detailed study of this and Łuczak [80] refined this analysis. The component structure of $\mathbb{G}_{n,m}$ for $m = n/2+o(n)$ is rather complicated and the proofs are technically challenging since those proofs require precise estimates of the number of very sparse connected labeled graphs. Nevertheless, it is possible to see that the largest component should be of order $n^{2/3}$ using a nice argument from Nachmias and Peres. They published a stronger version of this argument in [93].

Theorem 4.12 *Let $p = \frac{1}{n}$ and A be a large constant. Let Z be the size of the largest component in $\mathbb{G}_{n,p}$. Then*

$$(i) \quad \mathbb{P}\left(Z \leq \frac{1}{A} n^{2/3}\right) = O(A^{-1}),$$

$$(ii) \quad \mathbb{P}\left(Z \geq A n^{2/3}\right) = O(A^{-1}).$$

Proof We will prove part (i) of the theorem first. This is a standard application of the First Moment Method. Let X_k be the number of tree components of order k and let $k \in \left[\frac{1}{A} n^{2/3}, A n^{2/3}\right]$. Then, see also (4.11),

$$\mathbb{E}\, X_k = \binom{n}{k} k^{k-2} p^{k-1} (1-p)^{k(n-k)+\binom{k}{2}-k+1}.$$

But

$$\begin{aligned}(1-p)^{k(n-k)+\binom{k}{2}-k+1} &\sim (1-p)^{kn-k^2/2} \\ &= \exp\{(kn - k^2/2)\log(1-p)\} \\ &\sim \exp\left\{-\frac{kn - k^2/2}{n}\right\}.\end{aligned}$$

Hence, by the above and the approximation of the binomial for $k = o(n^{3/4})$ (see formula (2.12)), we have

$$\mathbb{E}\, X_k \sim \frac{n}{\sqrt{2\pi}\, k^{5/2}} \exp\left\{-\frac{k^3}{6n^2}\right\}. \tag{4.22}$$

So if

$$X = \sum_{\frac{1}{A} n^{2/3}}^{A n^{2/3}} X_k,$$

then

$$\begin{aligned}\mathbb{E}\, X &\sim \frac{1}{\sqrt{2\pi}} \int_{x=\frac{1}{A}}^{A} \frac{e^{-x^3/6}}{x^{5/2}} dx \\ &= \frac{4}{3\sqrt{\pi}} A^{3/2} + O(A^{1/2}).\end{aligned}$$

Arguing as in Lemma 4.8 we see that

$$\begin{aligned}\mathbb{E}\, X_k^2 &\leq \mathbb{E}\, X_k + (1+o(1))(\mathbb{E}\, X_k)^2, \\ \mathbb{E}(X_k X_l) &\leq (1+o(1))(\mathbb{E}\, X_k)(\mathbb{E}\, X_l), \quad k \neq l.\end{aligned}$$

It follows that

$$\mathbb{E}\, X^2 \leq \mathbb{E}\, X + (1+o(1))(\mathbb{E}\, X)^2.$$

Applying the Second Moment Method, we see that

$$\mathbb{P}(X > 0) \geq \frac{1}{(\mathbb{E}\, X)^{-1} + 1 + o(1)}$$

$$= 1 - O(A^{-1}),$$

which completes the proof of part (i).

To prove (ii) we first consider a breadth-first search (BFS) starting from, say, vertex x. We will use the notion of *stochastic dominance* (see Lemma 2.29).

We construct a sequence of sets $S_1 = \{x\}, S_2, \ldots$, where

$$S_{i+1} = \{v \notin S_i : \exists w \in S_i \text{ such that } (v, w) \in E(\mathbb{G}_{n,p})\}.$$

We have

$$\begin{aligned}\mathbb{E}(|S_{i+1}|\,|S_i) &\le (n - |S_i|)\left(1 - (1-p)^{|S_i|}\right)\\ &\le (n - |S_i|)|S_i|p\\ &\le |S_i|.\end{aligned}$$

So

$$\mathbb{E}\,|S_{i+1}| \le \mathbb{E}\,|S_i| \le \cdots \le \mathbb{E}\,|S_1| = 1. \tag{4.23}$$

We prove next that

$$\pi_k = \mathbb{P}(S_k \ne \emptyset) \le \frac{4}{k}. \tag{4.24}$$

This is clearly true for $k \le 4$, and we obtain (4.24) by induction from

$$\pi_{k+1} \le \sum_{i=1}^{n-1} \binom{n-1}{i} p^i (1-p)^{n-1-i} (1 - (1 - \pi_k)^i). \tag{4.25}$$

To explain the above inequality note that we can couple the construction of $S_1, S_2, \ldots, S_k$ with a (branching) process where $T_1 = \{1\}$ and T_{k+1} is obtained from T_k as follows: each T_k independently spawns $\text{Bin}(n-1, p)$ individuals. Note that $|T_k|$ stochastically dominates $|S_k|$. This is because in the BFS process, each $w \in S_k$ gives rise to at most $\text{Bin}(n-1, p)$ new vertices. Inequality (4.25) follows because $T_{k+1} \ne \emptyset$ implies that at least one of 1's children gives rise to descendants at level k. Going back to (4.25) we get

$$\begin{aligned}\pi_{k+1} &\le 1 - (1-p)^{n-1} - (1 - p + p(1 - \pi_k))^{n-1} + (1-p)^{n-1}\\ &= 1 - (1 - p\pi_k)^{n-1}\\ &\le \pi_k\left(1 - \frac{1}{4}\pi_k\right).\end{aligned} \tag{4.26}$$

This expression increases for $0 \le \pi_k \le 1$ and immediately gives $\pi_5 \le 3/4 \le 4/5$. In general, we have by induction that

$$\pi_{k+1} \le \frac{4}{k}\left(1 - \frac{1}{k}\right) \le \frac{4}{k+1},$$

completing the inductive proof of (4.24).

Let C_x be the component containing x and let $\rho_x = \max\{k : S_k \neq \emptyset\}$ in the BFS from x. Let

$$X = \left|\left\{x : |C_x| \geq n^{2/3}\right\}\right| \leq X_1 + X_2,$$

where

$$X_1 = \left|\left\{x : |C_x| \geq n^{2/3} \text{ and } \rho_x \leq n^{1/3}\right\}\right|,$$

$$X_2 = \left|\left\{x : \rho_x > n^{1/3}\right\}\right|.$$

It follows from (4.24) that

$$\mathbb{P}(\rho_x > n^{1/3}) \leq \frac{4}{n^{1/3}}$$

and so

$$\mathbb{E}\, X_2 \leq 4n^{2/3}.$$

Furthermore,

$$\begin{aligned}\mathbb{P}\left\{|C_x| \geq n^{2/3} \text{ and } \rho_x \leq n^{1/3}\right\} &\leq \mathbb{P}\left(|S_1| + \cdots + |S_{n^{1/3}}| \geq n^{2/3}\right)\\ &\leq \frac{1}{n^{1/3}},\end{aligned} \tag{4.27}$$

after using (4.23). So $\mathbb{E}\, X_1 \leq n^{2/3}$ and $\mathbb{E}\, X \leq 5n^{2/3}$.

Now let $C_{\max}$ denote the size of the largest component. Now

$$C_{\max} \leq |X| + n^{2/3},$$

where the addition of $n^{2/3}$ accounts for the case where $X = 0$.

So we have

$$\mathbb{E}\, C_{\max} \leq 6n^{2/3}$$

and part (ii) of the theorem follows from the Markov inequality (see Lemma 2.15).

□

Exercises

4.3.1. Let $C(k, k + l)$ denote the number of connected graphs with k vertices and $k + l$ edges, where $l = -1, 0, 1, 2, \ldots$ (e.g., $C(k, k - 1) = k^{k-2}$ is the number of labeled trees on k vertices). Let X be a random variable counting connected components of $\mathbb{G}_{n,p}$ with exactly k vertices and $k + l$ edges. Find the expected value of the random variable X for large n in terms of $C(k, k + l)$.

4.3.2. Verify equation (4.26).

4.3.3. Verify equation (4.27).

Problems for Chapter 4

4.1 Prove Theorem 4.11.

4.2 Suppose that $m = cn/2$, where $c > 1$ is a constant. Let C_1 denote the giant component of $\mathbb{G}_{n,m}$ assuming that it exists. Suppose that C_1 has $n' \leq n$ vertices and $m' \leq m$ edges. Let G_1, G_2 be two connected graphs with n' vertices from $[n]$ and m' edges. Show that

$$\mathbb{P}(C_1 = G_1) = \mathbb{P}(C_1 = G_2)$$

(i.e., C_1 is a uniformly random connected graph with n' vertices and m' edges).

4.3 Suppose that Z is the length of the cycle in a randomly chosen connected unicyclic graph on vertex set $[n]$. Show that

$$\mathbb{E}\,Z = \frac{n^{n-2}(N - n + 1)}{C(n,n)},$$

where $N = \binom{n}{2}$.

4.4 Suppose that $c < 1$. Show that w.h.p. the length of the longest path in $\mathbb{G}_{n,p}$, $p = \frac{c}{n}$ is $\sim \frac{\log n}{\log 1/c}$.

4.5 Suppose that $c \neq 1$ is constant. Show that w.h.p. the number of edges in the largest component that is a path in $\mathbb{G}_{n,p}$, $p = \frac{c}{n}$ is $\sim \frac{\log n}{c - \log c}$.

4.6 Let $p = \frac{1+\varepsilon}{n}$. Show that if ε is a small positive constant, then w.h.p. $\mathbb{G}_{n,p}$ contains a giant component of size $(2\varepsilon + O(\varepsilon^2))n$.

4.7 Let $m = \frac{n}{2} + s$, where $s = s(n) \geq 0$. Show that if $s \gg n^{2/3}$, then w.h.p. the random graph $\mathbb{G}_{n,m}$ contains exactly one complex component. (Recall that a component C is *complex* if it contains at least two distinct cycles. In terms of edges, C is complex if and only if it contains at last $|C| + 1$ edges.)

4.8 Let $m_k(n) = n(\log n + (k-1)\log\log n + \omega)/(2k)$, where $|\omega| \to \infty$, $|\omega| = o(\log n)$. Show that

$$\mathbb{P}(\mathbb{G}_{m_k} \not\supseteq k\text{-vertex tree component}) = \begin{cases} o(1) & \text{if } \omega \to -\infty, \\ 1 - o(1) & \text{if } \omega \to \infty. \end{cases}$$

4.9 Suppose that $p = \frac{c}{n}$, where $c > 1$ is a constant. Show that w.h.p. the giant component of $\mathbb{G}_{n,p}$ is nonplanar. (Hint: Assume that $c = 1 + \varepsilon$, where ε is small. Remove a few vertices from the giant so that the girth is large. Now use Euler's formula.)

4.10 Suppose that $p = c/n$, where $c > 1$ is constant and let $\beta = \beta(c)$ be the smallest root of the equation

$$\frac{1}{2}c\beta + (1-\beta)ce^{-c\beta} = \log\left(c(1-\beta)^{(\beta-1)/\beta}\right).$$

1. Show that if $\omega \to \infty$ and $\omega \leq k \leq \beta n$, then w.h.p. $\mathbb{G}_{n,p}$ contains no *maximal* induced tree of size k.
2. Show that w.h.p. $\mathbb{G}_{n,p}$ contains an induced tree of size $(\log n)^2$.
3. Deduce that w.h.p. $\mathbb{G}_{n,p}$ contains an induced tree of size at least βn.

4.11 Given a positive integer k, the *k-core of a graph* $G = (V, E)$ is the largest set $S \subseteq V$ such that the minimum degree in the vertex-induced subgraph $G[S]$ is at least k. Suppose that $c > 1$ and that $x < 1$ is the solution to $xe^{-x} = ce^{-c}$. Show that then w.h.p. the number of vertices in the 2-core C_2 of $G_{n,p}$, $p = c/n$ is asymptotically $\left(1 - \frac{x}{c}\right)^2 n$.

4.12 Let $k \geq 3$ be fixed and let $p = \frac{c}{n}$. Show that if c is sufficiently large, then w.h.p. the k-core of $\mathbb{G}_{n,p}$ is nonempty.

4.13 Let $k \geq 3$ be fixed and let $p = \frac{c}{n}$. Show that there exists $\theta = \theta(c, k) > 0$ such that w.h.p. all vertex sets S with $|S| \leq \theta n$ contain fewer than $k|S|/2$ edges. Deduce that w.h.p. either the k-core of $\mathbb{G}_{n,p}$ is empty or it has size at least θn.

4.14 Let $\mathbb{G}_{n,n,p}$ denote the random bipartite graph derived from the complete bipartite graph $K_{n,n}$ where each edge is included independently with probability p. Show that if $p = c/n$, where $c > 1$ is a constant, then w.h.p. $\mathbb{G}_{n,n,p}$ has a unique giant component of size $\sim 2G(c)n$, where $G(c)$ is as in Theorem 4.10.

4.15 Consider the bipartite random graph $\mathbb{G}_{n,n,p=c/n}$, with constant $c > 1$. Define $0 < x < 1$ to be the solution to $xe^{-x} = ce^{-c}$. Prove that w.h.p. the 2-core of $\mathbb{G}_{n,n,p=c/n}$ has $\sim 2(1 - x)\left(1 - \frac{x}{c}\right) n$ vertices and $\sim c\left(1 - \frac{x}{c}\right)^2 n$ edges.

5 Vertex Degrees

In this chapter we study some typical properties of the degree sequence of a random graph. We begin by discussing the typical degrees in a sparse random graph, i.e., one with $cn/2$ edges for some positive constant c. We prove some results on the asymptotic distribution of degrees. The average degree of a fixed vertex is c, and perhaps not surprisingly, the distribution of the degree of a fixed vertex is $Po(c)$, Poisson with mean c. The number of vertices Z_d of a given value d will then w.h.p. be concentrated around the mean $\mathbb{E}(Z_d)$, which will be approximately $nc^d e^{-c}/d!$. Hence, the number of vertices of degree d in a random graph is, roughly, $n/d^{d+o(d)}$ and so enjoys (super-)exponential decay as d grows. This is in sharp contrast to the polynomial decay observed in real-world networks, which we shall see later in Chapters 11 and 12.

We continue by looking at the typical values of the minimum and maximum degrees in dense random graphs, i.e., $G_{n,p}$ where p is constant. We find that the maximum and minimum degrees are w.h.p. equal to $np + O(\sqrt{n \log n})$. Surprisingly, even though the range of degrees is $O(\sqrt{n \log n})$, we find that there is a unique vertex of maximum degree. Further, this uniqueness spreads to the second largest and third largest and for a substantial range.

Given these properties of the degree sequence of dense graphs, we can then describe a simple canonical labeling algorithm that enables one to solve the graph isomorphism problem on a dense random graph.

5.1 Degrees of Sparse Random Graphs

Let us look first at the degree $\deg(v)$ of a vertex v in both models of random graphs. Observe that $\deg(v)$ in $\mathbb{G}_{n,p}$ is a binomially distributed random variable, with parameters $n-1$ and p, i.e., for $d = 0, 1, 2, \ldots, n-1$,

$$\mathbb{P}(\deg(v) = d) = \binom{n-1}{d} p^d (1-p)^{n-1-d},$$

while in $\mathbb{G}_{n,m}$ the distribution of $\deg(v)$ is hypergeometric, i.e.,

$$\mathbb{P}(\deg(v) = d) = \frac{\binom{n-1}{d}\binom{\binom{n-1}{2}}{m-d}}{\binom{\binom{n}{2}}{m}}.$$

It is important to observe that in both random graph models the degrees of different vertices are only mildly correlated.

In this section we concentrate on the number of vertices of a given degree when the random graph $\mathbb{G}_{n,p}$ is sparse, i.e., when the edge probability $p = o(1)$.

We return to the property that a random graph contains an isolated vertex (see Example 3.9, where we established the threshold for "disappearance" of such vertices from $\mathbb{G}_{n,p}$). To study this property more closely denote by $X_0 = X_{n,0}$ the number of isolated vertices in $\mathbb{G}_{n,p}$. Obviously,

$$\mathbb{E} X_0 = n(1-p)^{n-1},$$

and

$$\mathbb{E} X_0 \longrightarrow \begin{cases} \infty & \text{if } np - \log n \longrightarrow -\infty, \\ e^{-c} & \text{if } np - \log n \longrightarrow c, \quad c < \infty, \\ 0 & \text{if } np - \log n \longrightarrow \infty \end{cases} \tag{5.1}$$

as $n \longrightarrow \infty$.

To study the distribution of the random variable X_0 (and many other numerical characteristics of random graphs) as the number of vertices $n \longrightarrow \infty$, i.e., its asymptotic distribution, we apply a standard probabilistic technique based on moments.

Method of Moments Let X be a random variable with probability distribution completely determined by its moments. If $X_1, X_2, \ldots, X_n, \ldots$ are random variables with finite moments such that $\mathbb{E} X_n^k \longrightarrow \mathbb{E} X^k$ as $n \longrightarrow \infty$ for every integer $k \geq 1$, then the sequence of random variables $\{X_n\}$ converges in distribution to random variable X. The same is true if factorial moments of $\{X_n\}$ converge to factorial moments of X, i.e., when

$$\mathbb{E}(X_n)_k = E[X_n(X_n-1)\ldots(X_n-k+1)] \longrightarrow \mathbb{E}(X)_k \tag{5.2}$$

for every integer $k \geq 1$.

For further considerations we shall use the following notation.

We denote by $\mathrm{Po}(\lambda)$ the random variable with the Poisson distribution with parameter λ, while $N(0,1)$ denotes the variable with the standard normal distribution. We write $X_n \xrightarrow{D} X$ to say that a random variable X_n *converges in distribution* to a random variable X as $n \longrightarrow \infty$.

In random graphs we often count certain objects (vertex degrees, subgraphs of a given size, etc.). Then the respective random variables are sums of indicators I over a certain set of indices $\mathcal{I}$, i.e.,

$$S_n = \sum_{i \in \mathcal{I}} I_i.$$

Then the kth factorial moment counts the number of ordered k-tuples of objects with $I_i = 1$, i.e.,

$$(S_n)_k = \sum_{i_1,i_2,\ldots,i_k} I_{i_1} I_{i_2} \cdots I_{i_k},$$

where the summation is taken over all sequences of distinct indices $i_1, i_2, \ldots, i_k$.

The factorial moment variant of the Method of Moments is particularly useful when the limiting distribution is $\mathrm{Po}(\lambda)$, since the moments of a random variable with the Poisson distribution have a relatively complicated form while its kth factorial moment is λ^k. Therefore, proving convergence in the distribution of random variables such as S_n defined above we shall use the following lemma.

Lemma 5.1 *Let $S_n = \sum_{i \in \mathcal{I}} I_{n,i}$ be the sum of indicator random variables $I_{n,i}$. Suppose that there exists $\lambda \geq 0$ such that for every fixed $k \geq 1$,*

$$\lim_{n\to\infty} \mathbb{E}(S_n)_k = \sum_{i_1,i_2,\ldots,i_k} \mathbb{P}(I_{n,i_1} = 1, I_{n,i_2} = 1, \ldots, I_{n,i_k} = 1) = \lambda^k.$$

Then, for every $j \geq 0$,

$$\lim_{n\to\infty} \mathbb{P}(S_n = j) = e^{-\lambda} \frac{\lambda^j}{j!},$$

i.e., S_n converges in distribution to the Poisson distributed random variable with expectation λ ($S_n \xrightarrow{D} \mathrm{Po}(\lambda)$).

In the next theorem we discuss the asymptotic distribution of X_0 and claim that it passes through three phases: it starts in the normal phase; next when isolated vertices are close to "dying out," it moves through a Poisson phase; it finally ends up at the distribution concentrated at 0.

Theorem 5.2 *Let X_0 be the random variable counting isolated vertices in a random graph $\mathbb{G}_{n,p}$. Then, as $n \to \infty$,*

(i) $\tilde{X}_0 = (X_0 - \mathbb{E} X_0)/(\mathrm{Var}\, X_0)^{1/2} \xrightarrow{D} \mathrm{N}(0, 1)$ *if* $n^2 p \to \infty$ *and* $np - \log n \to -\infty$,

(ii) $X_0 \xrightarrow{D} \mathrm{Po}(e^{-c})$ *if* $np - \log n \to c,\ c < \infty$,

(iii) $X_0 \xrightarrow{D} 0$ *if* $np - \log n \to \infty$.

Proof For the proof of (i) we refer the reader to Chapter 6 of Janson, Łuczak, and Ruciński [66] (or to [14] and [74]). Note that statement (iii) has already been proved in Example 3.9.

To prove (ii) one has to show that if $p = p(n)$ is such that $np - \log n \to c$, then

$$\lim_{n\to\infty} \mathbb{P}(X_0 = k) = \frac{e^{-ck}}{k!} e^{-e^{-c}} \tag{5.3}$$

for $k = 0, 1, \ldots$. Now,

$$X_0 = \sum_{v \in V} I_v,$$

where

$$I_v = \begin{cases} 1 & \text{if } v \text{ is an isolated vertex in } \mathbb{G}_{n,p}, \\ 0 & \text{otherwise.} \end{cases}$$

So

$$\begin{aligned} \mathbb{E}\, X_0 = \sum_{v \in V} \mathbb{E}\, I_v &= n(1-p)^{n-1} = n \exp\{(n-1)\log(1-p)\} \\ &\sim e^{-c}. \end{aligned} \tag{5.4}$$

Statement (ii) follows from direct application of Lemma 5.1, since for every $k \geq 1$,

$$\mathbb{E}(X_0)_k = \sum_{i_1, i_2, \ldots, i_k} \mathbb{P}(I_{v_{i_1}} = 1, I_{v_{i_2}} = 1, \ldots, I_{v_{i_k}} = 1) = (n)_k (1-p)^{k(n-k) + \binom{k}{2}}.$$

Hence

$$\lim_{n \to \infty} \mathbb{E}(X_0)_k = e^{-ck},$$

i.e., X_0, when $p = (\log n + c)/n$, is asymptotically Poisson distributed with the expected value $\lambda = e^{-c}$. □

It follows immediately that for such an edge probability p,

$$\lim_{n \to \infty} \mathbb{P}(X_0 = 0) = e^{-e^{-c}}. \tag{5.5}$$

We next give a more general result describing the asymptotic distribution of the number $X_d = X_{n,d}$, $d \geq 1$ of vertices of any fixed degree d in a random graph. We shall see that the asymptotic behavior of X_d for $d \geq 1$ differs from the case $d = 0$ (isolated vertices). It is closely related to the asymptotic behavior of the expected value of X_d and the fact that the property of having a vertex of fixed degree d enjoys two thresholds, the first for its appearance and the second one for its disappearance (vertices of given degree are dying out when the edge probability p increases).

Theorem 5.3 *Let $X_d = X_{n,d}$ be the number of vertices of degree d, $d \geq 1$, in $\mathbb{G}_{n,p}$ and let $\lambda_1 = c^d/d!$ while $\lambda_2 = e^{-c}/d!$, where c is a constant. Then, as $n \to \infty$,*

(i) $X_d \xrightarrow{D} 0$ *if* $p \ll n^{-(d+1)/d}$,

(ii) $X_d \xrightarrow{D} \mathrm{Po}(\lambda_1)$ *if* $p \sim cn^{-(d+1)/d}$, $c < \infty$,

(iii) $\tilde{X}_d := (X_d - \mathbb{E}\, X_d)/(\mathrm{Var}\, X_d)^{1/2} \xrightarrow{D} \mathrm{N}(0, 1)$ *if* $p \gg n^{-(d+1)/d}$, *but* $pn - \log n - d \log\log n \to -\infty$,

(iv) $X_d \xrightarrow{D} \mathrm{Po}(\lambda_2)$ *if* $pn - \log n - d \log\log n \to c$, $-\infty < c < \infty$,

(v) $X_d \xrightarrow{D} 0$ *if* $pn - \log n - d \log\log n \to \infty$.

Proof The proofs of statements (i) and (v) are straightforward applications of the first moment method, while the proofs of (ii) and (iv) can be found in Chapter 3 of Bollobás [19] (see also Karoński and Ruciński [71] for estimates of the rate of convergence). The proof of (iii) can be found in [14]. □

The next theorem shows the concentration of X_d around its expectation when in $\mathbb{G}_{n,p}$ the edge probability $p = c/n$, i.e., when the average vertex degree is c.

Theorem 5.4 *Let $p = c/n$, where c is a constant. Let X_d denote the number of vertices of degree d in $\mathbb{G}_{n,p}$. Then, for $d = O(1)$, w.h.p.*

$$X_d \sim \frac{c^d e^{-c}}{d!} n.$$

Proof Assume that vertices of $\mathbb{G}_{n,p}$ are labeled $1, 2, \ldots, n$. We first compute $\mathbb{E}\, X_d$. Thus,

$$\begin{aligned}\mathbb{E}\, X_d = n\, \mathbb{P}(\deg(1) = d) &= n\binom{n-1}{d}\left(\frac{c}{n}\right)^d \left(1 - \frac{c}{n}\right)^{n-1-d} \\ &= n\frac{c^d e^{-c}}{d!}\left(1 + O\left(\frac{1}{n}\right)\right).\end{aligned} \tag{5.6}$$

We now compute the second moment. For this we need to estimate

$$\begin{aligned}&\mathbb{P}(\deg(1) = \deg(2) = d) \\ &\quad = \frac{c}{n}\left(\binom{n-2}{d-1}\left(\frac{c}{n}\right)^{d-1}\left(1-\frac{c}{n}\right)^{n-1-d}\right)^2 \\ &\qquad + \left(1 - \frac{c}{n}\right)\left(\binom{n-2}{d}\left(\frac{c}{n}\right)^{d}\left(1-\frac{c}{n}\right)^{n-2-d}\right)^2 \\ &\quad = \mathbb{P}(\deg(1) = d)\, \mathbb{P}(\deg(2) = d)\left(1 + O\left(\frac{1}{n}\right)\right).\end{aligned}$$

The first line here accounts for the case where $\{1, 2\}$ is an edge, and the second line deals with the case where it is not. Thus

$$\begin{aligned}\operatorname{Var} X_d &= \sum_{i=1}^{n}\sum_{j=1}^{n}\left[\mathbb{P}(\deg(i) = d, \deg(j) = d) - \mathbb{P}(\deg(1) = d)\, \mathbb{P}(\deg(2) = d)\right] \\ &\leq \sum_{i \neq j = 1}^{n} O\left(\frac{1}{n}\right) + \mathbb{E}\, X_d \leq An\end{aligned}$$

for some constant $A = A(c)$.

Applying the Chebyshev inequality (Lemma 2.18), we obtain

$$\mathbb{P}(|X_d - \mathbb{E}\, X_d| \geq tn^{1/2}) \leq \frac{A}{t^2},$$

which completes the proof. □

We conclude this section with a look at the asymptotic behavior of the maximum vertex degree when a random graph is sparse.

Theorem 5.5 *Let $\Delta(\mathbb{G}_{n,p})$ (resp. $\delta(\mathbb{G}_{n,p})$) denote the maximum (resp. minimum) degree of vertices of $\mathbb{G}_{n,p}$.*

(i) If $p = c/n$ for some constant $c > 0$, then w.h.p.

$$\Delta(\mathbb{G}_{n,p}) \sim \frac{\log n}{\log \log n}.$$

(ii) If $np = \omega \log n$, where $\omega \to \infty$, then w.h.p. $\delta(\mathbb{G}_{n,p}) \sim \Delta(\mathbb{G}_{n,p}) \sim np$.

Proof (i) Let $d_\pm = \left\lceil \frac{\log n}{\log\log n \pm 2 \log\log\log n} \right\rceil$. Then, if $d = d_-$,

$$\mathbb{P}(\exists v : \deg(v) \geq d) \leq n\binom{n-1}{d}\left(\frac{c}{n}\right)^d$$
$$\leq \exp\{\log n - d \log d + O(d)\}. \tag{5.7}$$

Let $\lambda = \frac{\log\log\log n}{\log\log n}$. Then

$$d \log d \geq \frac{\log n}{\log\log n} \cdot \frac{1}{1-2\lambda} \cdot (\log\log n - \log\log\log n + o(1))$$

$$= \frac{\log n}{\log\log n}(\log\log n + \log\log\log n + o(1)). \tag{5.8}$$

Plugging this into (5.7) shows that $\Delta(\mathbb{G}_{n,p}) \leq d_-$ w.h.p.

Now let $d = d_+$ and let X_d be the number of vertices of degree d in $\mathbb{G}_{n,p}$. Then

$$\mathbb{E}(X_d) = n\binom{n-1}{d}\left(\frac{c}{n}\right)^d \left(1-\frac{c}{n}\right)^{n-d-1} \to \infty. \tag{5.9}$$

Now, for vertices v, w, by the same argument as in the proof of Theorem 5.4, we have

$$\mathbb{P}(\deg(v) = \deg(w) = d) = (1+o(1))\,\mathbb{P}(\deg(v) = d)\,\mathbb{P}(\deg(w) = d),$$

and the Chebyshev inequality implies that $X_d > 0$ w.h.p. This completes the proof of (i).

Notice that statement (ii) is an easy consequence of the Chernoff bounds, Lemma 2.21. Let $\varepsilon = \omega^{-1/3}$. Then

$$\mathbb{P}(\exists v : |\deg(v) - np| \geq \varepsilon np) \leq 2ne^{-\varepsilon^2 np/3} = 2n^{-\omega^{1/3}/3} = o(n^{-1}).$$

$\square$

Exercises

5.1.1 Recall that the degree of a vertex in $\mathbb{G}_{n,p}$ has the binomial distribution $\text{Bin}(n-1, p)$. Hence,

$$\mathbb{E}\, X_d = n\binom{n-1}{d} p^d (1-p)^{n-1-d}.$$

Show that, as $n \to \infty$,

$$\mathbb{E}\, X_d \to \begin{cases} 0 & \text{if } p \ll n^{-(d+1)/d}, \\ \lambda_1 & \text{if } p \sim cn^{-(d+1)/d},\ c < \infty, \\ \infty & \text{if } p \gg n^{-(d+1)/d)} \text{ but} \\ & \quad pn - \log n - d \log\log n \to -\infty, \\ \lambda_2 & \text{if } pn - \log n - d \log\log n \to c,\ c < \infty, \\ 0 & \text{if } pn - \log n - d \log\log n \to \infty, \end{cases}$$

where

$$\lambda_1 = \frac{c^d}{d!} \quad \text{and} \quad \lambda_2 = \frac{e^{-c}}{d!}.$$

5.1.2 Suppose that $m = dn/2$, where d is constant. Prove that the number of vertices of degree k in $\mathbb{G}_{n,m}$ is asymptotically equal to $\frac{d^k e^{-d}}{k!} n$ for any fixed positive integer k.

5.1.3 Let $p = \frac{\log n + d \log\log n + c}{n}$, $d \geq 1$. Using the method of moments, prove that the number of vertices of degree d in $\mathbb{G}_{n,p}$ is asymptotically Poisson with mean $\frac{e^{-c}}{d!}$.

5.1.4 Prove parts (i) and (v) of Theorem 5.3.

5.1.5 Verify equation (5.4).

5.1.6 Verify equation (5.6).

5.1.7 Verify equation (5.7).

5.1.8 Verify equation (5.8).

5.1.9 Verify equation (5.9).

5.1.10 Show that if $p = O\left(\frac{\log n}{n}\right)$, then w.h.p. the maximum degree in $\mathbb{G}_{n,p}$ is $O(\log n)$.

5.2 Degrees of Dense Random Graphs

In this section we will concentrate on the case where the edge probability p is constant and see how the degree sequence can be used to solve the graph isomorphism problem w.h.p. The main result deals with the maximum vertex degree in a dense random graph and is instrumental in the solution of this problem.

Theorem 5.6 *Let $d_\pm = (n-1)p + (1 \pm \varepsilon)\sqrt{2(n-1)pq \log n}$, where $q = 1 - p$. If p is constant and $\varepsilon > 0$ is a small constant, then w.h.p.*

(i) $d_- \le \Delta(\mathbb{G}_{n,p}) \le d_+$.
(ii) There is a unique vertex of maximum degree.

Proof We break the proof of Theorem 5.6 into two lemmas.

Lemma 5.7 *Let* $d = (n-1)p + x\sqrt{(n-1)pq}$, *p be constant,* $x \le n^{1/3}(\log n)^2$, *where* $q = 1 - p$. *Then*

$$B_d = \binom{n-1}{d} p^d (1-p)^{n-1-d} = (1 + o(1))\sqrt{\frac{1}{2\pi npq}} e^{-x^2/2}.$$

Proof Stirling's formula gives

$$B_d = (1 + o(1))\sqrt{\frac{1}{2\pi npq}} \left(\left(\frac{(n-1)p}{d}\right)^{\frac{d}{n-1}} \left(\frac{(n-1)q}{n-1-d}\right)^{1-\frac{d}{n-1}} \right)^{n-1}. \tag{5.10}$$

Now

$$\begin{aligned}\left(\frac{d}{(n-1)p}\right)^{\frac{d}{n-1}} &= \left(1 + x\sqrt{\frac{q}{(n-1)p}}\right)^{\frac{d}{n-1}} \\ &= \exp\left\{x\sqrt{\frac{pq}{n-1}} + \frac{x^2 q}{2(n-1)} + O\left(\frac{x^3}{n^{3/2}}\right)\right\},\end{aligned} \tag{5.11}$$

whereas

$$\begin{aligned}\left(\frac{n-1-d}{(n-1)q}\right)^{1-\frac{d}{n-1}} &= \left(1 - x\sqrt{\frac{p}{(n-1)q}}\right)^{1-\frac{d}{n-1}} \\ &= \exp\left\{-x\sqrt{\frac{pq}{n-1}} + \frac{x^2 p}{2(n-1)} + O\left(\frac{x^3}{n^{3/2}}\right)\right\}.\end{aligned} \tag{5.12}$$

So

$$\left(\frac{d}{(n-1)p}\right)^{\frac{d}{n-1}} \left(\frac{n-1-d}{(n-1)q}\right)^{1-\frac{d}{n-1}} = \exp\left\{\frac{x^2}{2(n-1)} + O\left(\frac{x^3}{n^{3/2}}\right)\right\},$$

and the lemma follows from (5.10). □

The next lemma proves a strengthening of Theorem 5.6.

Lemma 5.8 *Let* $\varepsilon = 1/10$, *and p be constant and* $q = 1 - p$. *If*

$$d_\pm = (n-1)p + (1 \pm \varepsilon)\sqrt{2(n-1)pq\log n},$$

then w.h.p.

(i) $\Delta(\mathbb{G}_{n,p}) \le d_+$,
(ii) there are $\Omega(n^{2\varepsilon(1-\varepsilon)})$ *vertices of degree at least* d_-,
(iii) $\nexists\, u \ne v$ *such that* $\deg(u), \deg(v) \ge d_-$ *and* $|\deg(u) - \deg(v)| \le 10$.

Proof We first prove that as $x \to \infty$,

$$\frac{1}{x} e^{-x^2/2}\left(1 - \frac{1}{x^2}\right) \le \int_x^\infty e^{-y^2/2} dy \le \frac{1}{x} e^{-x^2/2}. \tag{5.13}$$

To see this notice

$$\int_x^\infty e^{-y^2/2} dy = -\int_x^\infty \frac{1}{y}\left(e^{-y^2/2}\right)' dy$$
$$= \frac{1}{x}e^{-x^2/2}\left(1-\frac{1}{x^2}\right) + O\left(\frac{1}{x^4}e^{-x^2/2}\right). \tag{5.14}$$

We can now prove statement (i). Let X_d be the number of vertices of degree d. Then $\mathbb{E}\, X_d = nB_d$ and so Lemma 5.7 implies that

$$\mathbb{E}\, X_d = (1+o(1))\sqrt{\frac{n}{2\pi pq}} \exp\left\{-\frac{1}{2}\left(\frac{d-(n-1)p}{\sqrt{(n-1)pq}}\right)^2\right\}$$

assuming that

$$d \le d_L = (n-1)p + (\log n)^2\sqrt{(n-1)pq}.$$

Also, if $d > (n-1)p$, then

$$\frac{B_{d+1}}{B_d} = \frac{(n-d-1)p}{(d+1)q} < 1,$$

and so if $d \ge d_L$,

$$\mathbb{E}\, X_d \le \mathbb{E}\, X_{d_L} \le n\exp\{-\Omega((\log n)^4)\}.$$

It follows that

$$\Delta(\mathbb{G}_{n,p}) \le d_L \quad w.h.p. \tag{5.15}$$

Now if $Y_d = X_d + X_{d+1} + \cdots + X_{d_L}$ for $d = d_\pm$, then

$$\begin{aligned}
\mathbb{E}\, Y_d &\sim \sum_{l=d}^{d_L} \sqrt{\frac{n}{2\pi pq}} \exp\left\{-\frac{1}{2}\left(\frac{l-(n-1)p}{\sqrt{(n-1)pq}}\right)^2\right\} \\
&\sim \sum_{l=d}^{\infty} \sqrt{\frac{n}{2\pi pq}} \exp\left\{-\frac{1}{2}\left(\frac{l-(n-1)p}{\sqrt{(n-1)pq}}\right)^2\right\} \\
&\sim \sqrt{\frac{n}{2\pi pq}} \int_{\lambda=d}^{\infty} \exp\left\{-\frac{1}{2}\left(\frac{\lambda-(n-1)p}{\sqrt{(n-1)pq}}\right)^2\right\} d\lambda.
\end{aligned} \tag{5.16}$$

The justification for (5.16) comes from

$$\sum_{l=d_L}^{\infty} \sqrt{\frac{n}{2\pi pq}} \exp\left\{-\frac{1}{2}\left(\frac{l-(n-1)p}{\sqrt{(n-1)pq}}\right)^2\right\}$$
$$= O(n) \sum_{x=(\log n)^2}^{\infty} e^{-x^2/2} = O(e^{-(\log n)^2/3}),$$

and

$$\sqrt{\frac{n}{2\pi pq}}\exp\left\{-\frac{1}{2}\left(\frac{d_+-(n-1)p}{\sqrt{(n-1)pq}}\right)^2\right\}=n^{-O(1)}.$$

If $d=(n-1)p+x\sqrt{(n-1)pq}$, then from (5.13) we have

$$\begin{aligned}\mathbb{E}Y_d &\sim \sqrt{\frac{n}{2\pi pq}}\int_{\lambda=d}^{\infty}\exp\left\{-\frac{1}{2}\left(\frac{\lambda-(n-1)p}{\sqrt{(n-1)pq}}\right)^2\right\}d\lambda\\ &=\sqrt{\frac{n}{2\pi pq}}\sqrt{(n-1)pq}\int_{y=x}^{\infty}e^{-y^2/2}dy\\ &\sim \frac{n}{\sqrt{2\pi}}\frac{1}{x}e^{-x^2/2}\\ &\begin{cases}\le n^{-2\varepsilon(1+\varepsilon)} & d=d_+,\\ \ge n^{2\varepsilon(1-\varepsilon)} & d=d_-.\end{cases}\end{aligned}\tag{5.17}$$

Part (i) follows from (5.17).

When $d=d_-$, we see from (5.17) that $\mathbb{E}Y_d\to\infty$. We use the second moment method to show that $Y_{d_-}\neq 0$ w.h.p.

$$\begin{aligned}\mathbb{E}(Y_d(Y_d-1)) &= n(n-1)\sum_{d\le d_1,d_2}^{d_L}\mathbb{P}(deg(1)=d_1,deg(2)=d_2)\\ &=n(n-1)\sum_{d\le d_1,d_2}^{d_L}[p\,\mathbb{P}(\hat{d}(1)=d_1-1,\hat{d}(2)=d_2-1)\\ &\quad+(1-p)\,\mathbb{P}(\hat{d}(1)=d_1,\hat{d}(2)=d_2)],\end{aligned}$$

where $\hat{d}(x)$ is the number of neighbors of x in $\{3,4,\dots,n\}$. Note that $\hat{d}(1)$ and $\hat{d}(2)$ are independent, and

$$\begin{aligned}\frac{\mathbb{P}(\hat{d}(1)=d_1-1)}{\mathbb{P}(\hat{d}(1)=d_1)}&=\frac{\binom{n-2}{d_1-1}(1-p)}{\binom{n-2}{d_1}p}=\frac{d_1(1-p)}{(n-1-d_1)p}\\ &=1+\tilde{O}(n^{-1/2}).\end{aligned}$$

In $\tilde{O}$ we ignore polylog factors.

Hence

$$\begin{aligned}&\mathbb{E}(Y_d(Y_d-1))\\ &=n(n-1)\sum_{d\le d_1,d_2}^{d_L}\left[\mathbb{P}(\hat{d}(1)=d_1)\,\mathbb{P}(\hat{d}(2)=d_2)(1+\tilde{O}(n^{-1/2}))\right]\\ &=n(n-1)\sum_{d\le d_1,d_2}^{d_L}\left[\mathbb{P}(deg(1)=d_1)\,\mathbb{P}(deg(2)=d_2)(1+\tilde{O}(n^{-1/2}))\right]\\ &=(\mathbb{E}Y_d)(\mathbb{E}Y_d-1)(1+\tilde{O}(n^{-1/2})),\end{aligned}$$

since

$$\frac{\mathbb{P}(\hat{d}(1) = d_1)}{\mathbb{P}(\deg(1) = d_1)} = \frac{\binom{n-2}{d_1}}{\binom{n-1}{d_1}}(1-p)^{-1}$$
$$= 1 + \tilde{O}(n^{-1/2}).$$

So, with $d = d_-$,

$$\mathbb{P}\left(Y_d \le \frac{1}{2}\mathbb{E}\,Y_d\right) \le \frac{\operatorname{Var} Y_d}{(\mathbb{E}\,Y_d)^2/4}$$
$$= \frac{\mathbb{E}(Y_d(Y_d-1)) + \mathbb{E}\,Y_d - (\mathbb{E}\,Y_d)^2}{(\mathbb{E}\,Y_d)^2/4}$$
$$= o(1). \tag{5.18}$$

This completes the proof of statement (ii). Finally,

$$\mathbb{P}(\neg(\text{iii})) \le o(1) + \binom{n}{2}\sum_{d_1=d_-}^{d_L}\sum_{|d_2-d_1|\le 10}\mathbb{P}(\deg(1)=d_1, \deg(2)=d_2)$$
$$= o(1) + \binom{n}{2}\sum_{d_1=d_-}^{d_L}\sum_{|d_2-d_1|\le 10}\Big[p\,\mathbb{P}(\hat{d}(1)=d_1-1)\,\mathbb{P}(\hat{d}(2)=d_2-1)$$
$$+ (1-p)\,\mathbb{P}(\hat{d}(1)=d_1)\,\mathbb{P}(\hat{d}(2)=d_2)\Big].$$

Now

$$\sum_{d_1=d_-}^{d_L}\sum_{|d_2-d_1|\le 10}\mathbb{P}(\hat{d}(1)=d_1-1)\,\mathbb{P}(\hat{d}(2)=d_2-1)$$
$$\le 21(1+\tilde{O}(n^{-1/2}))\sum_{d_1=d_-}^{d_L}\left[\mathbb{P}(\hat{d}(1)=d_1-1)\right]^2,$$

and by Lemma 5.7 and by (5.13) we have with

$$x = \frac{d_- - (n-1)p}{\sqrt{(n-1)pq}} \sim (1-\varepsilon)\sqrt{2\log n},$$

$$\sum_{d_1=d_-}^{d_L}\left[\mathbb{P}(\hat{d}(1)=d_1-1)\right]^2 \sim \frac{1}{2\pi pqn}\int_{y=x}^{\infty} e^{-y^2}\,dy$$
$$= \frac{1}{\sqrt{8}\pi pqn}\int_{z=x\sqrt{2}}^{\infty} e^{-z^2/2}\,dz \tag{5.19}$$
$$\sim \frac{1}{\sqrt{8}\pi pqn}\frac{1}{x\sqrt{2}}n^{-2(1-\varepsilon)^2}. \tag{5.20}$$

We get a similar bound for $\sum_{d_1=d_-}^{d_L} \sum_{|d_2-d_1|\leq 10} \left[\mathbb{P}(\hat{d}(1)=d_1\right]^2$. Thus

$$\begin{aligned}\mathbb{P}(\neg(\text{iii})) &= o\left(n^{2-1-2(1-\varepsilon)^2}\right)\\ &= o(1).\end{aligned}$$

□

Application to graph isomorphism

In this section we describe a procedure for *canonically labeling* a graph G. It is taken from Babai, Erdős, and Selkow [9]. If the procedure succeeds, then it is possible to quickly tell whether $G \cong H$ for any other graph H. (Here $\cong$ stands for isomorphic as graphs.)

Algorithm *LABEL*
Step 0: Input graph G and parameter L.
Step 1: Relabel the vertices of G so that they satisfy

$$\deg_G(v_1) \geq \deg_G(v_2) \geq \cdots \geq \deg_G(v_n).$$

If there exists $i < L$ such that $\deg_G(v_i) = \deg_G(v_{i+1})$, then **FAIL**.
Step 2: For $i > L$ let

$$X_i = \{j \in \{1,2,\ldots,L\} : \{v_i, v_j\} \in E(G)\}.$$

Relabel vertices $v_{L+1}, v_{L+2}, \ldots, v_n$ so that these sets satisfy

$$X_{L+1} \succ X_{L+2} \succ \cdots \succ X_n$$

where $\succ$ denotes lexicographic order.
If there exists $i < n$ such that $X_i = X_{i+1}$, then **FAIL**.

Suppose now that the above ordering/labeling procedure LABEL succeeds for G. Given an n vertex graph H, we run LABEL on H.

(i) If LABEL fails on H, then $G \not\cong H$.
(ii) Suppose that the ordering generated on $V(H)$ is $w_1, w_2, \ldots, w_n$. Then

$$G \cong H \Leftrightarrow v_i \to w_i \text{ is an isomorphism.}$$

It is straightforward to verify (i) and (ii) for large n.

Theorem 5.9 *Let p be a fixed constant, $q = 1 - p$, and let $\rho = p^2 + q^2$ and let $L = 3\log_{1/\rho} n$. Then w.h.p. LABEL succeeds on $\mathbb{G}_{n,p}$.*

Proof Part (iii) of Lemma 5.8 implies that Step 1 succeeds w.h.p. We must now show that w.h.p. $X_i \neq X_j$ for all $i \neq j > L$. There is a slight problem because the edges from $v_i, i > L$ to $v_j, j \leq L$ are conditioned by the fact that the latter vertices are those of highest degree.

Now fix i, j and let $\hat{G} = \mathbb{G}_{n,p} \setminus \{v_i, v_j\}$. It follows from Lemma 5.8 that if $i, j > L$, then w.h.p. the L largest degree vertices of $\hat{G}$ and $\mathbb{G}_{n,p}$ coincide. So, w.h.p., we can compute X_i, X_j with respect to $\hat{G}$ to create $\hat{X}_i, \hat{X}_j$, which are independent of the edges incident with v_i, v_j. It follows that if $i, j > L$, then $\hat{X}_i = X_i$ and $\hat{X}_j = X_j$, and this avoids our conditioning problem. Denote by $N_{\hat{G}}(v)$ the set of the neighbors of vertex v in graph $\hat{G}$. Then

$$
\begin{aligned}
&\mathbb{P}(\text{Step 2 fails}) \\
&\quad \leq o(1) + \mathbb{P}(\exists v_i, v_j : N_{\hat{G}}(v_i) \cap \{v_1, \dots, v_L\} = N_{\hat{G}}(v_j) \cap \{v_1, \dots, v_L\}) \\
&\quad \leq o(1) + \binom{n}{2}(p^2 + q^2)^L \\
&\quad = o(1).
\end{aligned}
$$

□

Corollary 5.10 *If $0 < p < 1$ is constant, then w.h.p. $G_{n,p}$ has a unique automorphism, i.e., the identity automorphism. (An automorphism of a graph $G = (V, E)$ is a map $\varphi : V \to V$ such that $\{x, y\} \in E$ if and only if $\{\varphi(x), \varphi(y)\} \in E$.)*

See Problem 5.3.

Application to edge coloring

The *chromatic index* $\chi'(G)$ of a graph G is the minimum number of colors that can be used to color the edges of G so that if two edges share a vertex, then they have a different color. Vizing's theorem states that

$$
\Delta(G) \leq \chi'(G) \leq \Delta(G) + 1.
$$

Also, if there is a unique vertex of maximum degree, then $\chi'(G) = \Delta(G)$. So, it follows from Theorem 5.6 (ii) that, for constant p, w.h.p. we have $\chi'(\mathbb{G}_{n,p}) = \Delta(\mathbb{G}_{n,p})$.

Exercises

5.2.1 Verify equation (5.10).

5.2.2 Verify equation (5.11).

5.2.3 Verify equation (5.12).

5.2.4 Verify equation (5.14).

5.2.5 Verify equation (5.18).

Problems for Chapter 5

5.1 Suppose that $c > 1$ and that $x < 1$ is the solution to $xe^{-x} = ce^{-c}$. Show that if $c = O(1)$ is fixed, then w.h.p. the giant component of $\mathbb{G}_{n,p}, p = \frac{c}{n}$ has $\sim \frac{c^k e^{-c}}{k!} \left(1 - \left(\frac{x}{c}\right)^k\right) n$ vertices of degree $k \geq 1$.

5.2 Show that if $0 < p < 1$ is constant, then w.h.p. the minimum degree δ in $\mathbb{G}_{n,p}$ satisfies

$$|\delta - (n-1)q - \sqrt{2(n-1)pq \log n}| \leq \varepsilon\sqrt{2(n-1)pq \log n},$$

where $q = 1 - p$ and $\varepsilon = 1/10$.

5.3 Use the canonical labeling of Theorem 5.9 to show that w.h.p. $G_{n,1/2}$ has exactly one automorphism, the identity automorphism.

6 Connectivity

Whether a graph is connected, i.e., there is a path between any two of its vertices, is of particular importance. Therefore, in this chapter we first establish the threshold for the connectivity of a random graph. We then view this property in terms of the graph process and show that w.h.p. the random graph becomes connected at precisely the time when the last isolated vertex joins the giant component. This "hitting time" result is the pre-cursor to several similar results.

After this, we deal with k-connectivity, i.e., the parameter which measures the strength of connectivity of a graph. We show that the threshold for this property is the same as for the existence of vertices of degree k in a random graph. In fact, a much stronger statement, similar to that for connectedness, can be proved. Namely, that a random graph becomes k-connected as soon as the last vertex of degree $k-1$ disappears.

In general, one can observe in many results from Part I of the text that one of the characteristic features of Erdős–Rényi–Gilbert random graphs, trivial graph-theoretic necessary conditions, such as minimum degree $\delta(G) > 0$ for connectedness, or $\delta(G) > k-1$ for k-connectivity, becomes sufficient w.h.p.

6.1 Connectivity

The first result of this chapter is from Erdős and Rényi [42].

Theorem 6.1 *Let* $m = \frac{1}{2}n(\log n + c_n)$. *Then*

$$\lim_{n\to\infty} \mathbb{P}(\mathbb{G}_m \text{ is connected}) = \begin{cases} 0 & \text{if } c_n \to -\infty, \\ e^{-e^{-c}} & \text{if } c_n \to c \text{ (constant)}, \\ 1 & \text{if } c_n \to \infty. \end{cases}$$

Proof To prove the theorem we consider, as before, the random graph $\mathbb{G}_{n,p}$. It suffices to prove that, when $p = \frac{\log n + c}{n}$,

$$\mathbb{P}(\mathbb{G}_{n,p} \text{ is connected }) \to e^{-e^{-c}}.$$

We use Theorem 3.4 to translate to $\mathbb{G}_m$ and then use (3.6) and monotonicity for $c_n \to \pm\infty$.

Let $X_k = X_{k,n}$ be the number of components with k vertices in $\mathbb{G}_{n,p}$ and consider the complement of the event that $\mathbb{G}_{n,p}$ is connected. Then

$$\begin{aligned}\mathbb{P}(\mathbb{G}_{n,p} \text{ is } \textbf{not} \text{ connected}) &= \mathbb{P}\left(\bigcup_{k=1}^{n/2}(\mathbb{G}_{n,p} \text{ has a component of order } k)\right)\\ &= \mathbb{P}\left(\bigcup_{k=1}^{n/2}\{X_k > 0\}\right).\end{aligned}$$

Note that X_1 counts here isolated vertices and therefore

$$\mathbb{P}(X_1 > 0) \leq \mathbb{P}(\mathbb{G}_{n,p} \text{ is } \textbf{not} \text{ connected }) \leq \mathbb{P}(X_1 > 0) + \sum_{k=2}^{n/2} \mathbb{P}(X_k > 0).$$

Now

$$\sum_{k=2}^{n/2} \mathbb{P}(X_k > 0) \leq \sum_{k=2}^{n/2} \mathbb{E}\, X_k \leq \sum_{k=2}^{n/2} \binom{n}{k} k^{k-2} p^{k-1} (1-p)^{k(n-k)} = \sum_{k=2}^{n/2} u_k.$$

Now, for $2 \leq k \leq 10$,

$$\begin{aligned}u_k &\leq e^k n^k \left(\frac{\log n + c}{n}\right)^{k-1} e^{-k(n-10)\frac{\log n + c}{n}}\\ &\leq (1+o(1)) e^{k(1-c)} \left(\frac{\log n}{n}\right)^{k-1},\end{aligned}$$

and for $k > 10$,

$$\begin{aligned}u_k &\leq \left(\frac{ne}{k}\right)^k k^{k-2} \left(\frac{\log n + c}{n}\right)^{k-1} e^{-k(\log n + c)/2}\\ &\leq n \left(\frac{e^{1-c/2+o(1)} \log n}{n^{1/2}}\right)^k.\end{aligned}$$

So

$$\sum_{k=2}^{n/2} u_k = O\left(n^{o(1)-1}\right). \tag{6.1}$$

It follows that

$$\mathbb{P}(\mathbb{G}_{n,p} \text{ is connected }) = \mathbb{P}(X_1 = 0) + o(1).$$

But we already know (see Theorem 5.2) that for $p = (\log n + c)/n$ the number of isolated vertices in $\mathbb{G}_{n,p}$ has an asymptotically Poisson distribution and therefore, as in (5.5),

$$\lim_{n\to\infty} \mathbb{P}(X_1 = 0) = e^{-e^{-c}},$$

and so the theorem follows. □

It is possible to tweak the proof of Theorem 6.1 to give a more precise result stating that a random graph becomes connected exactly at the moment when the last isolated vertex disappears.

Theorem 6.2 *Consider the random graph process $\{\mathbb{G}_m\}$. Let*

$$m_1^* = \min\{m : \delta(\mathbb{G}_m) \geq 1\},$$

$$m_c^* = \min\{m : \mathbb{G}_m \text{ is connected}\}.$$

Then, w.h.p.

$$m_1^* = m_c^*.$$

Proof Let

$$m_\pm = \frac{1}{2} n \log n \pm \frac{1}{2} n \log \log n$$

and

$$p_\pm = \frac{m_\pm}{N} \sim \frac{\log n \pm \log \log n}{n},$$

where $N = \binom{n}{2}$.

We first show that w.h.p.

(i) G_{m_-} consists of a giant connected component plus a set V_1 of at most $2 \log n$ isolated vertices,

(ii) G_{m_+} is connected.

Assume (i) and (ii). It follows that w.h.p.

$$m_- \leq m_1^* \leq m_c^* \leq m_+.$$

Since $\mathbb{G}_{m_-}$ consists of a connected component and a set of isolated vertices V_1, to create $\mathbb{G}_{m_+}$ we add $m_+ - m_-$ random edges. Note that $m_1^* = m_c^*$ if none of these edges is contained in V_1. Thus

$$\begin{aligned} \mathbb{P}(m_1^* < m_c^*) &\leq o(1) + (m_+ - m_-) \frac{\frac{1}{2}|V_1|^2}{N - m_+} \\ &\leq o(1) + \frac{2n((\log n)^2) \log \log n}{\frac{1}{2}n^2 - O(n \log n)} \\ &= o(1). \end{aligned}$$

Thus to prove the theorem, it is sufficient to verify (i) and (ii).

Let

$$p_- = \frac{m_-}{N} \sim \frac{\log n - \log \log n}{n},$$

and let X_1 be the number of isolated vertices in $\mathbb{G}_{n,p_-}$. Then

$$\mathbb{E} X_1 = n(1 - p_-)^{n-1}$$

$$\sim ne^{-np_-}$$
$$\sim \log n.$$

Moreover,

$$\mathbb{E}\, X_1^2 = \mathbb{E}\, X_1 + n(n-1)(1-p_-)^{2n-3}$$
$$\leq \mathbb{E}\, X_1 + (\mathbb{E}\, X_1)^2(1-p_-)^{-1}.$$

So,

$$\operatorname{Var} X_1 \leq \mathbb{E}\, X_1 + 2(\mathbb{E}\, X_1)^2 p_-$$

and

$$\mathbb{P}(X_1 \geq 2\log n) = \mathbb{P}(|X_1 - \mathbb{E}\, X_1| \geq (1+o(1))\,\mathbb{E}\, X_1)$$
$$\leq (1+o(1))\left(\frac{1}{\mathbb{E}\, X_1} + 2p_-\right)$$
$$= o(1).$$

Having at least $2\log n$ isolated vertices is a monotone property, and so w.h.p. $\mathbb{G}_{m_-}$ has less than $2\log n$ isolated vertices.

To show that the rest of $\mathbb{G}_m$ is a single connected component we let X_k, $2 \leq k \leq n/2$ be the number of components with k vertices in $\mathbb{G}_{p_-}$. Repeating the calculations for p_- from the proof of Theorem 6.1, we have

$$\mathbb{E}\left(\sum_{k=2}^{n/2} X_k\right) = O\left(n^{o(1)-1}\right).$$

Let

$$\mathcal{E} = \{\exists \text{ component of order } 2 \leq k \leq n/2\}.$$

Then

$$\mathbb{P}(\mathbb{G}_{m_-} \in \mathcal{E}) \leq O(\sqrt{n})\,\mathbb{P}(\mathbb{G}_{n,p_-} \in \mathcal{E})$$
$$= o(1),$$

and this completes the proof of (i).

To prove (ii) (that G_{m_+} is connected w.h.p.) we note that (ii) follows from the fact that $\mathbb{G}_{n,p}$ is connected w.h.p. for $np - \log n \to \infty$ (see Theorem 6.1). By implication $\mathbb{G}_m$ is connected w.h.p. if $n\frac{m}{N} - \log n \to \infty$. But

$$\frac{nm_+}{N} \sim \log n + \log\log n. \tag{6.2}$$

□

Exercises

6.1.1 Verify equation (6.1).

6.1.2 Verify equation (6.2).

6.2 k-Connectivity

In this section we show that the threshold for the existence of vertices of degree k is also the threshold for the k-connectivity of a random graph. Recall that a graph G is k-connected if the removal of at most $k-1$ vertices of G does not disconnect it. In light of the previous result it should be expected that a random graph becomes k-connected as soon as the last vertex of degree $k-1$ disappears. This is true and follows from the results of Erdős and Rényi [44]. Here is a weaker statement. The stronger statement is left as an exercise, Exercise 6.1.

Theorem 6.3 *Let* $m = \frac{1}{2}n\left(\log n + (k-1)\log\log n + c_n\right)$, $k = 1, 2, \ldots$. *Then*

$$\lim_{n\to\infty} \mathbb{P}(\mathbb{G}_m \text{ is } k\text{-connected}) = \begin{cases} 0 & \text{if } c_n \to -\infty, \\ e^{-\frac{e^{-c}}{(k-1)!}} & \text{if } c_n \to c, \\ 1 & \text{if } c_n \to \infty. \end{cases}$$

Proof Let

$$p = \frac{\log n + (k-1)\log\log n + c}{n}.$$

We will prove that, in $\mathbb{G}_{n,p}$, with edge probability p above,

(i) the expected number of vertices of degree at most $k-2$ is $o(1)$,
(ii) the expected number of vertices of degree $k-1$ is approximately $\frac{e^{-c}}{(k-1)!}$.

We have

$$\begin{aligned} &\mathbb{E}(\text{number of vertices of degree } t \le k-1) \\ &\qquad = n\binom{n-1}{t} p^t (1-p)^{n-1-t} \sim n\frac{n^t}{t!}\frac{(\log n)^t}{n^t}\frac{e^{-c}}{n(\log n)^{k-1}}, \end{aligned}$$

and (i) and (ii) follow immediately, see Exercises 6.2.1 and 6.2.2.

The distribution of the number of vertices of degree $k-1$ is asymptotically Poisson, as may be verified by the Method of Moments (see Exercise 5.1.3).

We now show that, if

$$\mathcal{A}(S,T) = \left\{T \text{ is a component of } \mathbb{G}_{n,p} \setminus S\right\},$$

then

$$\mathbb{P}\left(\exists S, T,\ |S| < k,\ 2 \le |T| \le \frac{1}{2}(n - |S|) : \mathcal{A}(S,T)\right) = o(1).$$

This implies that if $\delta(\mathbb{G}_{n,p}) \ge k$, then $\mathbb{G}_{n,p}$ is k-connected and Theorem 6.3 follows. $|T| \ge 2$ because if $T = \{v\}$, then v has degree less than k.

We can assume that S is minimal and then the neighborhood $N(T) = S$ and denote $s = |S|$, $t = |T|$. T is connected, and so it contains a tree with $t-1$ edges. Also each vertex of S is incident with an edge from S to T, and so there are at least s edges between S and T. Thus, if $p = (1+o(1))\frac{\log n}{n}$, then

$$\begin{aligned}
\mathbb{P}(\exists S,T) &\leq o(1) + \sum_{s=1}^{k-1}\sum_{t=2}^{(n-s)/2}\binom{n}{s}\binom{n}{t}t^{t-2}p^{t-1}\binom{st}{s}p^s(1-p)^{t(n-s-t)} \\
&\leq p^{-1}\sum_{s=1}^{k-1}\sum_{t=2}^{(n-s)/2}\left(\frac{ne}{s}\cdot(te)\cdot p\cdot e^{tp}\right)^s\left(ne\cdot p\cdot e^{-(n-t)p}\right)^t \\
&\leq p^{-1}\sum_{s=1}^{k-1}\sum_{t=2}^{(n-s)/2} A^t B^s, \qquad (6.3)
\end{aligned}$$

where

$$\begin{aligned}
A &= nepe^{-(n-t)p} = e^{1+o(1)}n^{-1+(t+o(t))/n}\log n, \\
B &= ne^2tpe^{tp} = e^{2+o(1)}tn^{(t+o(t))/n}\log n.
\end{aligned}$$

Now if $2 \leq t \leq \log n$, then $A = n^{-1+o(1)}$ and $B = O((\log n)^2)$. On the other hand, if $t > \log n$, then we can use $A \leq n^{-1/3}$ and $B \leq n^2$ to see that the sum in (6.3) is $o(1)$. □

Exercises

6.2.1 Verify statement (i) of Theorem 6.3.

6.2.2 Verify statement (ii) of Theorem 6.3.

Problems for Chapter 6

6.1 Let $k = O(1)$ and let m_k^* be the hitting time for minimum degree at least k in the graph process. Let t_k^* be the hitting time for k-connectivity. Show that $m_k^* = t_k^*$ w.h.p.

6.2 Let $m = m_1^*$ be as in Theorem 6.2 and let $e_m = (u,v)$, where u has degree 1. Let $0 < c < 1$ be a positive constant. Show that w.h.p. there is no triangle containing vertex v.

6.3 Let $m = m_1^*$ as in Theorem 6.2 and let $e_m = (u,v)$, where u has degree 1. Let $0 < c < 1$ be a positive constant. Show that w.h.p. the degree of v in G_m is at least $c\log n$.

6.4 Suppose that $n\log n \ll m \leq n^{3/2}$ and let $d = 2m/n$. Let $S_i(v)$ be the set of vertices at distance i from vertex v. Show that w.h.p. $|S_i(v)| \geq (d/2)^i$ for all $v \in [n]$ and $1 \leq i \leq \frac{2\log n}{3\log d}$.

6.5 Suppose that $m \gg n\log n$ and let $d = m/n$. Using the previous question, show that w.h.p. there are at least $d/2$ internally vertex disjoint paths of length at most $\frac{4\log n}{3\log d}$ between any pair of vertices in $G_{n,m}$.

6.6 Suppose that $m \gg n \log n$ and let $d = m/n$. Suppose that we randomly color the edges of $G_{n,m}$ with q colors where $q \gg \frac{(\log n)^2}{(\log d)^2}$. Show that w.h.p. there is a *rainbow* path between every pair of vertices. (A path is a rainbow if each of its edges has a different color.)

6.7 Let $\mathbb{G}_{n,n,p}$ be the random bipartite graph with vertex bi-partition $V = (A, B)$, $A = [1, n], B = [n + 1, 2n]$, in which each of the n^2 possible edges appears independently with probability p. Let $p = \frac{\log n + \omega}{n}$, where $\omega \to \infty$. Show that w.h.p. $\mathbb{G}_{n,n,p}$ is connected.

6.8 Show that for every $\varepsilon > 0$ there exists $c_\varepsilon > 0$ such that the following is true w.h.p. If $c \geq c_\varepsilon$ and $p = c/n$ and we remove any set of at most $(1 - \varepsilon)cn/2$ edges from $\mathbb{G}_{n,p}$, then the remaining graph contains a component of size at least $\varepsilon n/4$.

6.9 Show that $\mathbb{P}(\mathbb{G}_{n,p} \text{ is connected}) \leq (1 - (1 - p)^{n-1})^{n-1}$.

6.10 Show that the expected number $\mathbb{E}(c_{n,k})$ of components on k vertices in $\mathbb{G}_{n,p}$ can be bounded from above

$$\mathbb{E}(c_{n,k}) \leq \frac{1}{1 - q^k} \mathbb{P}(\text{Bin}(n, 1 - q^k) = k),$$

where $q = 1 - p$.

7 Small Subgraphs

Graph theory is replete with theorems stating conditions for the existence of a subgraph H in a larger graph G. For example Turán's theorem [110] states that a graph with n vertices and more than $\left(1 - \frac{1}{r}\right) \frac{n^2}{2}$ edges must contain a copy of K_{r+1}. In this chapter we see instead how many random edges are required to have a particular fixed size subgraph w.h.p. In addition, we will consider the distribution of the number of copies of strictly balanced subgraphs.

From these general results, one can deduce thresholds for small trees, stars, cliques, bipartite cliques, and many other small subgraphs which play an important role in the analysis of the properties not only of classic random graphs but also in the interpretation of characteristic features of real-world networks.

Computing the frequency of small subgraphs is a fundamental problem in network analysis, used across diverse domains: bioinformatics, social sciences, and infrastructure networks studies. The high frequencies of certain subgraphs in real networks give a quantifiable method of proving they are not Erdős–Rényi. The distributions of small subgraphs are used to evaluate network models, summarize real networks, and classify vertex roles, among other things.

7.1 Thresholds

In this section we will look for a threshold for the appearance of any fixed graph H, with $v_H = |V(H)|$ vertices and $e_H = |E(H)|$ edges. The property that a random graph contains H as a subgraph is clearly monotone increasing. It is also transparent that "denser" graphs appear in a random graph "later" than "sparser" ones. More precisely, denote by

$$d(H) = \frac{e_H}{v_H}, \tag{7.1}$$

the *density* of a graph H. Notice that $2d(H)$ is the average vertex degree in H. We begin with the analysis of the asymptotic behavior of the expected number of copies of H in the random graph $\mathbb{G}_{n,p}$.

Lemma 7.1 *Let X_H denote the number of copies of H in $\mathbb{G}_{n,p}$,*

$$\mathbb{E}\, X_H = \binom{n}{v_H} \frac{v_H!}{\mathrm{aut}(H)} p^{e_H},$$

where aut(H) *is the number of automorphisms of H.*

Proof The complete graph on n vertices K_n contains $\binom{n}{v_H} a_H$ distinct copies of H, where a_H is the number of copies of H in K_{v_H}. Thus

$$\mathbb{E}\, X_H = \binom{n}{v_H} a_H p^{e_H},$$

and all we need to show is that

$$a_H \times \mathrm{aut}(H) = v_H!.$$

Each permutation σ of $[v_H] = \{1, 2, \ldots, v_H\}$ defines a unique copy of H as follows: A copy of H corresponds to a set of e_H edges of K_{v_H}. The copy H_σ corresponding to σ has edges $\{(x_{\sigma(i)}, y_{\sigma(i)}) : 1 \le i \le e_H\}$, where $\{(x_j, y_j) : 1 \le j \le e_H\}$ is some fixed copy of H in K_{v_H}. But $H_\sigma = H_{\tau\sigma}$ if and only if for each i there is j such that $(x_{\tau\sigma(i)}, y_{\tau\sigma(i)}) = (x_{\sigma(j)}, y_{\sigma(j)})$, i.e., if τ is an automorphism of H. □

Theorem 7.2 *Let H be a fixed graph with $e_H > 0$. Suppose $p = o\left(n^{-1/d(H)}\right)$. Then w.h.p. $\mathbb{G}_{n,p}$ contains no copies of H.*

Proof Suppose that $p = \omega^{-1} n^{-1/d(H)}$, where $\omega = \omega(n) \to \infty$ as $n \to \infty$. Then

$$\mathbb{E}\, X_H = \binom{n}{v_H} \frac{v_H!}{\mathrm{aut}(H)} p^{e_H} \le n^{v_H} \omega^{-e_H} n^{-e_H/d(H)} = \omega^{-e_H}.$$

Thus

$$\mathbb{P}(X_H > 0) \le \mathbb{E}\, X_H \to 0 \text{ as } n \to \infty.$$

□

From our previous experience one would expect that when $\mathbb{E}\, X_H \to \infty$ as $n \to \infty$ the random graph $\mathbb{G}_{n,p}$ would contain H as a subgraph w.h.p. Let us check whether such a phenomenon also holds in this case. So consider the case when $pn^{1/d(H)} \to \infty$, where $p = \omega n^{-1/d(H)}$ and $\omega = \omega(n) \to \infty$ as $n \to \infty$. Then for some constant $c_H > 0$,

$$\mathbb{E}\, X_H \ge c_H n^{v_H} \omega^{e_H} n^{-e_H/d(H)} = c_H \omega^{e_H} \to \infty.$$

However, as we will see, this is not always enough for $\mathbb{G}_{n,p}$ to contain a copy of a given graph H w.h.p. To see this, consider the graph H given in Figure 7.1.

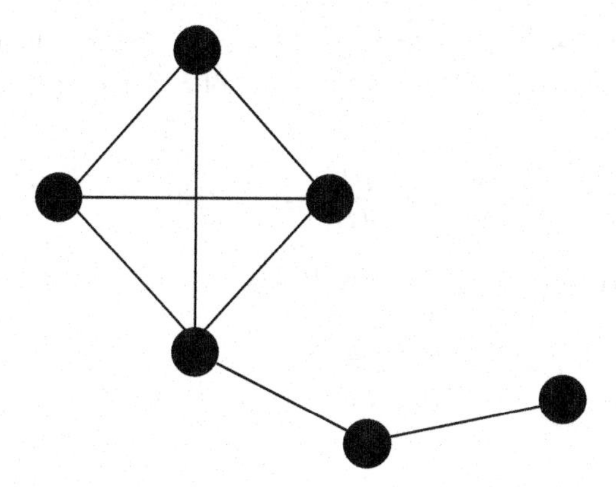

Figure 7.1 A kite

Here $v_H = 6$ and $e_H = 8$. Let $p = n^{-5/7}$. Now $1/d(H) = 6/8 > 5/7$ and so

$$\mathbb{E}\, X_H \sim c_H n^{6-8\times 5/7} \to \infty.$$

On the other hand, if $\hat{H} = K_4$, then

$$\mathbb{E}\, X_{\hat{H}} \leq n^{4-6\times 5/7} \to 0,$$

and so w.h.p. there are no copies of $\hat{H}$ and hence no copies of H.

The reason for such "strange" behavior is quite simple. Our graph H is in fact *not balanced*, since its overall density is smaller than the density of one of its subgraphs, i.e., of $\hat{H} = K_4$. So we need to introduce another density characteristic of graphs, namely the maximum subgraph density defined as follows:

$$m(H) = \max\{d(K) : K \subseteq H\}. \tag{7.2}$$

A graph H is *balanced* if $m(H) = d(H)$. It is *strictly balanced* if $d(H) > d(K)$ for all proper subgraphs $K \subset H$.

Now we are ready to determine the threshold for the existence of a copy of H in $\mathbb{G}_{n,p}$. Erdős and Rényi [43] proved this result for balanced graphs. The threshold for any graph H was first found by Bollobás in [19], and an alternative, deterministic argument to derive the threshold was presented in [70]. A simple proof, given here, is due to Ruciński and Vince [103].

Theorem 7.3 *Let H be a fixed graph with $e_H > 0$. Then*

$$\lim_{n\to\infty} \mathbb{P}(H \subseteq \mathbb{G}_{n,p}) = \begin{cases} 0 & \textit{if } pn^{1/m(H)} \to 0, \\ 1 & \textit{if } pn^{1/m(H)} \to \infty. \end{cases}$$

Proof Let $\omega = \omega(n) \to \infty$ as $n \to \infty$. The first statement follows from Theorem 7.2. Notice that if we choose $\hat{H}$ to be a subgraph of H with $d(\hat{H}) = m(H)$ (such a subgraph always exists since we do not exclude $\hat{H} = H$), then $p = \omega^{-1} n^{-1/d(\hat{H})}$ implies that $\mathbb{E}\, X_{\hat{H}} \to 0$. Therefore, w.h.p. $\mathbb{G}_{n,p}$ contains no copies of $\hat{H}$, and so it does not contain H as well.

To prove the second statement we use the Second Moment Method. Suppose now that $p = \omega n^{-1/m(H)}$. Denote by $H_1, H_2, \ldots, H_t$ all copies of H in the complete graph on $\{1, 2, \ldots, n\}$. Note that

$$t = \binom{n}{v_H} \frac{v_H!}{\text{aut}(H)}, \tag{7.3}$$

where aut(H) is the number of automorphisms of H. For $i = 1, 2, \ldots, t$ let

$$I_i = \begin{cases} 1 & \text{if } H_i \subseteq \mathbb{G}_{n,p}, \\ 0 & \text{otherwise.} \end{cases}$$

Let $X_H = \sum_{i=1}^t I_i$. Then

$$\begin{aligned} \operatorname{Var} X_H &= \sum_{i=1}^t \sum_{j=1}^t \operatorname{Cov}(I_i, I_j) = \sum_{i=1}^t \sum_{j=1}^t (\mathbb{E}(I_i I_j) - (\mathbb{E}\, I_i)(\mathbb{E}\, I_j)) \\ &= \sum_{i=1}^t \sum_{j=1}^t (\mathbb{P}(I_i = 1, I_j = 1) - \mathbb{P}(I_i = 1)\, \mathbb{P}(I_j = 1)) \\ &= \sum_{i=1}^t \sum_{j=1}^t \left(\mathbb{P}(I_i = 1, I_j = 1) - p^{2e_H}\right). \end{aligned}$$

Observe that random variables I_i and I_j are independent if and only if H_i and H_j are edge disjoint. In this case $\operatorname{Cov}(I_i, I_j) = 0$ and such terms vanish from the above summation. Therefore, we consider only pairs (H_i, H_j) with $H_i \cap H_j = K$ for some graph K with $e_K > 0$. So,

$$\begin{aligned} \operatorname{Var} X_H &= O\left(\sum_{K \subseteq H, e_K > 0} n^{2v_H - v_K} \left(p^{2e_H - e_K} - p^{2e_H}\right)\right) \\ &= O\left(n^{2v_H} p^{2e_H} \sum_{K \subseteq H, e_K > 0} n^{-v_K} p^{-e_K}\right). \end{aligned}$$

On the other hand,

$$\mathbb{E}\, X_H = \binom{n}{v_H} \frac{v_H!}{\text{aut}(H)} p^{e_H} = \Omega\left(n^{v_H} p^{e_H}\right).$$

Thus, by the Second Moment Method,

$$\begin{aligned} \mathbb{P}(X_H = 0) \le \frac{\operatorname{Var} X_H}{(\mathbb{E}\, X_H)^2} &= O\left(\sum_{K \subseteq H, e_K > 0} n^{-v_K} p^{-e_K} \right) \\ &= O\left(\sum_{K \subseteq H, e_K > 0} \left(\frac{1}{\omega n^{1/d(K) - 1/m(H)}} \right)^{e_K} \right) \\ &= o(1). \end{aligned}$$

Hence w.h.p., the random graph $\mathbb{G}_{n,p}$ contains a copy of the subgraph H when $pn^{1/m(H)} \to \infty$. □

Exercises

7.1.1 Draw a graph which is (a) balanced but not strictly balanced, (b) unbalanced.

7.1.2 Are the small graphs listed below, balanced or unbalanced? (a) a tree, (b) a cycle, (c) a complete graph, (d) a regular graph, (e) the Petersen graph, (f) a graph composed of a complete graph on four vertices and a triangle, sharing exactly one vertex.

7.1.3 Determine (directly, not from the statement of Theorem 7.3) thresholds $\hat{p}$ for $\mathbb{G}_{n,p} \supseteq G$, for graphs listed in the previous exercise. Do the same for the thresholds of G in $\mathbb{G}_{n,m}$.

7.2 Asymptotic Distributions

We will now study the asymptotic distribution of the number X_H of copies of a fixed graph H in $\mathbb{G}_{n,p}$. We start at the threshold, so we assume that $np^{m(H)} \to c,\ c > 0$, where $m(H)$ denotes, as before, the maximum subgraph density of H. Now, if H is not balanced, i.e., its maximum subgraph density exceeds the density of H, then $\mathbb{E}\, X_H \to \infty$ as $n \to \infty$, and one can show that there is a sequence of numbers a_n, increasing with n, such that the asymptotic distribution of X_H/a_n coincides with the distribution of a random variable counting the number of copies of a subgraph K of H for which $m(H) = d(K)$. Note that K is itself a balanced graph. However, the asymptotic distribution of balanced graphs on the threshold, although computable, cannot be given in a closed form. The situation changes dramatically if we assume that the graph H whose copies in $\mathbb{G}_{n,p}$ we want to count is *strictly balanced*, i.e., when for every proper subgraph K of H, $d(K) < d(H) = m(H)$. The following result is due to Bollobás [19] and Karoński and Ruciński [69].

Theorem 7.4 *If H is a strictly balanced graph and $np^{m(H)} \to c, c > 0$, then $X_H \xrightarrow{D} \mathrm{Po}(\lambda)$, as $n \to \infty$, where $\lambda = c^{v_H}/aut(H)$.*

Proof Denote, as before, by $H_1, H_2, \ldots, H_t$ all copies of H in the complete graph on $\{1, 2, \ldots, n\}$. For $i = 1, 2, \ldots, t$, let

$$I_{H_i} = \begin{cases} 1 & \text{if } H_i \subseteq \mathbb{G}_{n,p}, \\ 0 & \text{otherwise.} \end{cases}$$

Then $X_H = \sum_{i=1}^{t} I_{H_i}$ and the kth factorial moment of X_H, $k = 1, 2 \ldots$,

$$\mathbb{E}(X_H)_k = \mathbb{E}[X_H(X_H - 1) \cdots (X_H - k + 1)],$$

can be written as

$$\begin{aligned} \mathbb{E}(X_H)_k &= \sum_{i_1, i_2, \ldots, i_k} \mathbb{P}(I_{H_{i_1}} = 1, I_{H_{i_2}} = 1, \ldots, I_{H_{i_k}} = 1) \\ &= D_k + \overline{D}_k, \end{aligned}$$

where the summation is taken over all k-element sequences of distinct indices i_j from $\{1, 2, \ldots, t\}$, while D_k and $\overline{D}_k$ denote the partial sums taken over all (ordered) k tuples of copies of H which are, respectively, pairwise vertex disjoint (D_k) and not all pairwise vertex disjoint ($\overline{D}_k$). Now, observe that

$$D_k = \sum_{i_1, i_2, \ldots, i_k} \mathbb{P}(I_{H_{i_1}} = 1)\,\mathbb{P}(I_{H_{i_2}} = 1) \cdots \mathbb{P}(I_{H_{i_k}} = 1)$$
$$= \binom{n}{v_H, v_H, \ldots, v_H} (a_H p^{e_H})^k$$
$$\sim (\mathbb{E}\, X_H)^k.$$

So assuming that $np^{d(H)} = np^{m(H)} \to c$ as $n \to \infty$,

$$D_k \sim \left(\frac{c^{v_H}}{\mathrm{aut}(H)}\right)^k. \tag{7.4}$$

On the other hand, we will show that

$$\overline{D}_k \to 0 \text{ as } n \to \infty. \tag{7.5}$$

Consider the family $\mathcal{F}_k$ of all (mutually nonisomorphic) graphs obtained by taking unions of k not all pairwise vertex disjoint copies of the graph H. Suppose $F \in \mathcal{F}_k$ has v_F vertices ($v_H \le v_F \le kv_H - 1$) and e_F edges, and let $d(F) = e_F/v_F$ be its density. To prove that (7.5) holds we need the following lemma.

Lemma 7.5 *If $F \in \mathcal{F}_k$, then $d(F) > m(H)$.*

Proof Define

$$f_F = m(H)v_F - e_F. \tag{7.6}$$

We will show (by induction on $k \ge 2$) that $f_F < 0$ for all $F \in \mathcal{F}_k$. First note that $f_H = 0$ and that $f_K > 0$ for every proper subgraph K of H, since H is strictly balanced. Notice also that the function f is modular, i.e., for any two graphs F_1 and F_2,

$$f_{F_1 \cup F_2} = f_{F_1} + f_{F_2} - f_{F_1 \cap F_2}. \tag{7.7}$$

Assume that the copies of H composing F are numbered in such a way that $H_{i_1} \cap H_{i_2} \ne \emptyset$. If $F = H_{i_1} \cup H_{i_2}$, then (7.6) and $f_{H_1} = f_{H_2} = 0$ imply

$$f_{H_{i_1} \cup H_{i_2}} = -f_{H_{i_1} \cap H_{i_2}} < 0.$$

For arbitrary $k \ge 3$, let $F' = \bigcup_{j=1}^{k-1} H_{i_j}$ and $K = F' \cap H_{i_k}$. Then by the inductive assumption we have $f_{F'} < 0$ while $f_K \ge 0$ since K is a subgraph of H (in extreme cases K can be H itself or an empty graph). Therefore,

$$f_F = f_{F'} + f_{H_{i_k}} - f_K = f_{F'} - f_K < 0,$$

which completes the induction and implies that $d(F) > m(H)$. □

Let C_F be the number of sequences $H_{i_1}, H_{i_2}, \ldots, H_{i_k}$ of k distinct copies of H such that

$$V\Big(\bigcup_{j=1}^{k} H_{i_j}\Big) = \{1, 2, \ldots, v_F\} \text{ and } \bigcup_{j=1}^{k} H_{i_j} \cong F.$$

Then, by Lemma 7.5,

$$\overline{D}_k = \sum_{F \in \mathcal{F}_k} \binom{n}{v_F} C_F\, p^{e_F} = O(n^{v_F} p^{e_F})$$
$$= O\left(\left(np^{d(F)}\right)^{v(F)}\right) = o(1),$$

and so (7.5) holds. Summarizing,

$$\mathbb{E}(X_H)_k \sim \left(\frac{c^{v_H}}{\operatorname{aut}(H)}\right)^k,$$

and the theorem follows by Lemma 5.1. □

The following theorem describes the asymptotic behavior of the number of copies of a graph H in $\mathbb{G}_{n,p}$ past the threshold for the existence of a copy of H. It holds regardless of whether or not H is balanced or strictly balanced. We state the theorem but we do not supply a proof (see Ruciński [102]).

Theorem 7.6 *Let H be a fixed (nonempty) graph. If $np^{m(H)} \to \infty$ and $n^2(1-p) \to \infty$, then $(X_H - \mathbb{E}\, X_H)/(\operatorname{Var} X_H)^{1/2} \xrightarrow{D} \mathrm{N}(0, 1)$ as $n \to \infty$*

Exercises

7.2.1 Let f_F be a graph function defined as

$$f_F = a\, v_F + b\, e_F,$$

where a, b are constants, while v_F and e_F denote, respectively, the number of vertices and edges of a graph F. Show that the function f_F is modular.

7.2.2 Let X_e be the number of isolated edges in $\mathbb{G}_{n,p}$. Determine when the random variable X_e has asymptotically the Poisson distribution.

7.2.3 Determine (directly, not applying Theorem 7.4) when the random variable counting the number of copies of a triangle in $\mathbb{G}_{n,p}$ has asymptotically the Poisson distribution.

Problems for Chapter 7

7.1 For a graph G a *balanced extension* of G is a graph F such that $G \subseteq F$ and $m(F) = d(F) = m(G)$. Applying the result of Győri, Rothschild, and Ruciński [58] that every graph has a balanced extension, deduce Bollobás's result (Theorem 7.3) from that of Erdős and Rényi (threshold for balanced graphs).

7.2 Let H be a fixed graph and let $p = cn^{-1/m(H)}$, where $c > 0$ is a constant. Show that w.h.p. all copies of H in $\mathbb{G}_{n,p}$ are *induced* copies of H. (A copy of H is induced in a host graph G if G restricted to the vertices of the copy is H, i.e., no extra edges.)

7.3 Let H be a fixed graph and let $p = cn^{-1/m(H)}$, where $c > 0$ is a constant. Show that w.h.p. no two copies of H in $\mathbb{G}_{n,p}$ are within distance 10 of each other.

7.4 Let $\ell \geq 3$ be fixed. Show that if $n^{\ell-2}p^{\binom{\ell}{2}} \gg \log n$, then w.h.p. every edge of $\mathbb{G}_{n,p}$ is contained in a copy of K_ℓ.

7.5 Suppose that $0 < p < 1$ is constant. Show with the aid of McDiarmid's inequality (see Lemma 9.6) that the number of triangles Z in $\mathbb{G}_{n,p}$ satisfies $Z \sim n^3p^2/6$ w.h.p.

8 Large Subgraphs

The previous chapter dealt with the existence of small subgraphs of a fixed size. In this chapter we concern ourselves with the existence of large subgraphs, most notably perfect matchings and Hamilton cycles. The famous theorems of Hall and Tutte give necessary and sufficient conditions for a bipartite and an arbitrary graph respectively to contain a perfect matching. Hall's theorem, in particular, can be used to establish that the threshold for having a perfect matching in a random bipartite graph can be identified with that of having no isolated vertices.

Having dealt with perfect matchings, we turn our attention to long paths in sparse random graphs, i.e., in those where we expect a linear number of edges and show that, under such circumstances, a random graph contains a cycle of length $\Omega(n)$ w.h.p.

We next study one of the most celebrated and difficult problems in the first 10 years after the publication of the seminal Erdős and Rényi paper on the evolution of random graphs: the existence of a Hamilton cycle in a random graph, the question left open in [43]. The solution can be credited to Hungarian mathematicians: Ajtai, Komlós and Szemerédi, and Bollobás. We establish the precise limiting probability that $\mathbb{G}_{n,p}$ contains a Hamilton cycle. This is equal to the limiting probability that $\mathbb{G}_{n,p}$ has minimum degree 2. It means that a trivial necessary condition for a graph being Hamiltonian, i.e., $\delta(G) > 1$, is also sufficient for $\mathbb{G}_{n,p}$ w.h.p.

In the last section of this chapter we consider the general problem of the existence of arbitrary spanning subgraphs H in a random graph, where we bound the maximum degree $\Delta(H)$.

8.1 Perfect Matchings

Before we move to the problem of the existence of a perfect matching, i.e., a collection of independent edges covering all of the vertices of a graph, in our main object of study, the random graph $\mathbb{G}_{n,p}$, we will analyze the same problem in a random bipartite graph. This problem is much simpler than the respective one for $\mathbb{G}_{n,p}$ and provides a general approach to finding a perfect matching in a random graph.

Bipartite Graphs

Let $\mathbb{G}_{n,n,p}$ be the random bipartite graph with vertex bi-partition $V = (A, B)$, $A = [1, n]$, $B = [n + 1, 2n]$, in which each of the n^2 possible edges appears independently with probability p. The following theorem was first proved by Erdős and Rényi [45].

Theorem 8.1 *Let $\omega = \omega(n)$, $c > 0$ be a constant, and $p = \frac{\log n+\omega}{n}$. Then*

$$\lim_{n\to\infty} \mathbb{P}(\mathbb{G}_{n,n,p} \text{ has a perfect matching}) = \begin{cases} 0 & \text{if } \omega \to -\infty, \\ e^{-2e^{-c}} & \text{if } \omega \to c, \\ 1 & \text{if } \omega \to \infty. \end{cases}$$

Moreover,

$$\lim_{n\to\infty} \mathbb{P}(\mathbb{G}_{n,n,p} \text{ has a perfect matching}) = \lim_{n\to\infty} \mathbb{P}(\delta(\mathbb{G}_{n,n,p}) \geq 1).$$

Proof We will use Hall's condition for the existence of a perfect matching in a bipartite graph. It states that a bipartite graph contains a perfect matching if and only if the following condition is satisfied:

$$\forall S \subseteq A,\ |N(S)| \geq |S|, \tag{8.1}$$

where for a set of vertices S, $N(S)$ denotes the set of neighbors of S.

It is convenient to replace (8.1) by

$$\forall S \subseteq A,\ |S| \leq \frac{n}{2},\ \ |N(S)| \geq |S|, \tag{8.2}$$

$$\forall T \subseteq B,\ |T| \leq \frac{n}{2},\ \ |N(T)| \geq |T|. \tag{8.3}$$

This is because if $|S| > n/2$ and $|N(S)| < |S|$, then $T = B \setminus N(S)$ will violate (8.3).

Now we can restrict our attention to S, T satisfying (a) $|S| = |T| + 1$ and (b) each vertex in T has at least two neighbors in S. Take a pair S, T with $|S| + |T|$ as small as possible. If the minimum degree $\delta \geq 1$, then $|S| \geq 2$.

(i) If $|S| > |T| + 1$, we can remove $|S| - |T| - 1$ vertices from $|S|$ – contradiction.

(ii) Suppose $\exists w \in T$ such that w has less than two neighbors in S. Remove w and its (unique) neighbor in $|S|$-contradiction.

It follows that

$$\begin{aligned}
&\mathbb{P}(\exists v : v \text{ is isolated}) \leq \mathbb{P}(\nexists \text{ a perfect matching}) \\
&\leq \mathbb{P}(\exists v : v \text{ is isolated}) + 2\,\mathbb{P}(\exists S \subseteq A, T \subseteq B, 2 \leq k = |S| \leq n/2, \\
&\qquad\qquad |T| = k - 1, N(S) \subseteq T \text{ and } e(S : T) \geq 2k - 2).
\end{aligned}$$

Here $e(S : T)$ denotes the number of edges between S and T, and $e(S : T)$ can be assumed to be at least $2k - 2$ because of (b) above.

Suppose now that $p = \frac{\log n + c}{n}$ for some constant c. Then let Y denote the number of sets S and T not satisfying conditions (8.2) and (8.3). Then

$$\begin{aligned}
\mathbb{E}\,Y &\le 2\sum_{k=2}^{n/2} \binom{n}{k}\binom{n}{k-1}\binom{k(k-1)}{2k-2} p^{2k-2}(1-p)^{k(n-k)} \\
&\le 2\sum_{k=2}^{n/2} \left(\frac{ne}{k}\right)^k \left(\frac{ne}{k-1}\right)^{k-1} \left(\frac{ke(\log n + c)}{2n}\right)^{2k-2} e^{-npk(1-k/n)} \\
&\le \sum_{k=2}^{n/2} n \left(\frac{e^{O(1)} n^{k/n} (\log n)^2}{n}\right)^k \\
&= \sum_{k=2}^{n/2} u_k.
\end{aligned}$$

Case 1: $2 \le k \le n^{3/4}$.

$$u_k \le n((e^{O(1)} n^{-1} \log n)^2)^k.$$

So

$$\sum_{k=2}^{n^{3/4}} u_k = O\left(\frac{1}{n^{1-o(1)}}\right).$$

Case 2: $n^{3/4} < k \le n/2$.

$$u_k \le n^{1-k(1/2-o(1))}.$$

So

$$\sum_{n^{3/4}}^{n/2} u_k = O\left(n^{-n^{3/4}/3}\right)$$

and

$$\mathbb{P}(\nexists \text{ a perfect matching}) = \mathbb{P}(\exists \text{ isolated vertex}) + o(1).$$

Let X_0 denote the number of isolated vertices in $\mathbb{G}_{n,n,p}$. Then

$$\mathbb{E}\,X_0 = 2n(1-p)^n \sim 2e^{-c}.$$

It follows in fact via inclusion-exclusion or the Method of Moments that we have

$$\mathbb{P}(X_0 = 0) \sim e^{-2e^{-c}}. \tag{8.4}$$

To prove the case for $|\omega| \to \infty$ we can use monotonicity and (3.6) and the fact that $e^{-e^{-2c}} \to 0$ if $c \to -\infty$ and $e^{-e^{-2c}} \to 1$ if $c \to \infty$. □

Nonbipartite Graphs

We now consider $\mathbb{G}_{n,p}$. We could try to replace Hall's theorem by Tutte's theorem. A proof along these lines was given by Erdős and Rényi [46]. We can however get away with a simpler approach based on the expansion properties of $\mathbb{G}_{n,p}$. The proof here can be traced back to Bollobás and Frieze [25].

We write $\mathbb{G}_{n,p}$ as the union of two independent copies $\mathbb{G}_1 \cup \mathbb{G}_2$. We show that w.h.p. small sets in $\mathbb{G}_1$ have many neighbors. Then we show that if there is a maximum size matching in a graph $\mathbb{G}$ that does not cover vertex v, then there is a set of vertices $A(v)$ such that the addition of any $\{v, w\}$, $w \in A(v)$ to $\mathbb{G}$ would increase the size of the largest matching in $\mathbb{G}$. This set has few neighbors in $\mathbb{G}$ and so w.h.p. it is large in $\mathbb{G}_1$. This implies that there are $\Omega(n^2)$ edges (*boosters*) of the form $\{x, y\}$ where $y \in A(x)$ whose addition increases the maximum matching size. This is the role of $\mathbb{G}_2$.

Theorem 8.2 *Let $\omega = \omega(n)$, $c > 0$ be a constant, and let $p = \frac{\log n + c_n}{n}$. Then*

$$\lim_{\substack{n\to\infty \\ n \text{ even}}} \mathbb{P}(\mathbb{G}_{n,p} \text{ has a perfect matching}) = \begin{cases} 0 & \text{if } c_n \to -\infty, \\ e^{-e^{-c}} & \text{if } c_n \to c, \\ 1 & \text{if } c_n \to \infty. \end{cases}$$

Moreover,

$$\lim_{n\to\infty} \mathbb{P}(\mathbb{G}_{n,p} \text{ has a perfect matching}) = \lim_{n\to\infty} \mathbb{P}(\delta(\mathbb{G}_{n,p}) \geq 1).$$

Proof We will for convenience only consider the case where $c_n = \omega \to \infty$ and $\omega = o(\log n)$. If $c_n \to -\infty$, then there are isolated vertices w.h.p. and our proof can easily be modified to handle the case $c_n \to c$.

Our combinatorial tool that replaces Tutte's theorem is the following: We say that a matching M *isolates* a vertex v if no edge of M contains v.

For a graph G we let

$$\mu(G) = \max \{|M| : M \text{ is a matching in } G\}. \tag{8.5}$$

Let $G = (V, E)$ be a graph without a perfect matching, i.e., $\mu(G) < \lfloor |V|/2 \rfloor$. Fix $v \in V$ and suppose that M is a maximum matching that isolates v. Let $S_0(v, M) = \{u \neq v : M \text{ isolates } u\}$. If $u \in S_0(v, M)$ and $e = \{x, y\} \in M$ and $f = \{u, x\} \in E$, then *flipping* e, f replaces M by $M' = M + f - e$. Here e is *flipped-out*. Note that $y \in S_0(v, M')$.

Now fix a maximum matching M that isolates v and let

$$A(v, M) = \bigcup_{M'} S_0(v, M')$$

where we take the union over M' obtained from M by a sequence of flips.

Lemma 8.3 *Let G be a graph without a perfect matching and let M be a maximum matching and v be a vertex isolated by M. Then $|N_G(A(v, M))| < |A(v, M)|$.*

Proof Suppose that $x \in N_G(A(v,M))$ and that $f = \{u,x\} \in E$, where $u \in A(v,M)$. Now there exists y such that $e = \{x,y\} \in M$, else $x \in S_0(v,M) \subseteq A(v,M)$. We claim that $y \in A(v,M)$, and this will prove the lemma. Since then, every neighbor of $A(v,M)$ is the neighbor via an edge of M.

Suppose that $y \notin A(v,M)$. Let M' be a maximum matching that (i) isolates u and (ii) is obtainable from M by a sequence of flips. Now $e \in M'$ because if e has been flipped out, then either x or y is placed in $A(v,M)$. But then we can do another flip with M', e and the edge $f = \{u,x\}$, placing $y \in A(v,M)$, contradiction. □

We now change notation and write $A(v)$ in place of $A(v,M)$, understanding that there is some maximum matching that isolates v. Note that if $u \in A(v)$, then there is some maximum matching that isolates u, and so $A(u)$ is well defined. Furthermore, it is always the case that if v is isolated by some maximum matching and $u \in A(v)$, then $\mu(G + \{u,v\}) = \mu(G) + 1$.

Now let

$$p = \frac{\log n + \theta \log\log n + \omega}{n},$$

where $\theta \geq 0$ is a fixed integer and $\omega \to \infty$ and $\omega = o(\log\log n)$.

We have introduced θ so that we can use some of the following results for the Hamilton cycle problem.

We write

$$\mathbb{G}_{n,p} = \mathbb{G}_{n,p_1} \cup \mathbb{G}_{n,p_2},$$

where

$$p_1 = \frac{\log n + \theta \log\log n + \omega/2}{n} \tag{8.6}$$

and

$$1 - p = (1 - p_1)(1 - p_2) \text{ so that } p_2 \sim \frac{\omega}{2n}. \tag{8.7}$$

Note that Theorem 6.3 implies:

$$\text{The minimum degree in } \mathbb{G}_{n,p_1} \text{ is at least } \theta + 1 \ w.h.p. \tag{8.8}$$

We consider a process where we add the edges of $\mathbb{G}_{n,p_2}$ one at a time to $\mathbb{G}_{n,p_1}$. We want to argue that if the current graph does not have a perfect matching, then there is a good chance that adding such an edge $\{x,y\}$ will increase the size of a largest matching. This will happen if $y \in A(x)$. If we know that w.h.p. every set S for which $|N_{\mathbb{G}_{n,p_1}}(S)| < |S|$ satisfies $|S| \geq \alpha n$ for some constant $\alpha > 0$, then

$$\mathbb{P}(y \in A(x)) \geq \frac{\binom{\alpha n}{2} - O(n)}{\binom{n}{2}} \geq \frac{\alpha^2}{2}, \tag{8.9}$$

provided we have only looked at $O(n)$ edges of $\mathbb{G}_{n,p_2}$ so far.

This is because the edges we add will be uniformly random and there will be at least $\binom{\alpha n}{2}$ edges $\{x,y\}$ where $y \in A(x)$. Here given an initial x we can include edges $\{x',y'\}$ where $x' \in A(x)$ and $y' \in A(x')$.

In the light of this we now argue that sets S, with $|N_{\mathbb{G}_{n,p_1}}(S)| < (1+\theta)|S|$, are w.h.p. of size $\Omega(n)$.

Lemma 8.4 *Let $K = 100(\theta + 7)$. W.h.p. $S \subseteq [n], |S| \leq \frac{n}{2e(\theta+5)}$ implies $|N(S)| \geq (\theta + 1)|S|$, where $N(S) = N_{\mathbb{G}_{n,p_1}}(S)$.*

Proof Let a vertex of graph $\mathbb{G}_{n,p_1}$ be large if its degree is at least $\lambda = \frac{\log n}{100}$, and small otherwise. Denote by $LARGE$ and $SMALL$ the set of large and small vertices in $\mathbb{G}_{n,p_1}$, respectively.

***Claim* 1** *W.h.p. if $v, w \in SMALL$, then $dist(v, w) \geq 5$.*

Proof If v, w are small and connected by a short path P, then v, w will have few neighbors outside P, and conditional on P existing, v having few neighbors outside P is independent of w having few neighbors outside P. Hence,

$$
\begin{aligned}
&\mathbb{P}(\exists v, w \in SMALL \text{ in } \mathbb{G}_{n,p_1} \text{ such that } dist(v, w) < 5) \\
&\leq \binom{n}{2} \left(\sum_{l=0}^{3} n^l p_1^{l+1} \right) \left(\sum_{k=0}^{\lambda} \binom{n}{k} p_1^k (1 - p_1)^{n-k-5} \right)^2 \\
&\leq n^2 \cdot \frac{n^3 (\log n)^4}{n^4} \left(\sum_{k=0}^{\lambda} \frac{(\log n)^k}{k!} \cdot \frac{(\log n)^{(\theta+1)/100} \cdot e^{-\omega/2}}{n \log n} \right)^2 \\
&\leq 2n (\log n)^4 \left(\frac{(\log n)^\lambda}{\lambda!} \cdot \frac{(\log n)^{(\theta+1)/100} \cdot e^{-\omega/2}}{n \log n} \right)^2 \qquad (8.10) \\
&= O\left(\frac{(\log n)^{O(1)}}{n} (100e)^{\frac{2 \log n}{100}} \right) \\
&= O(n^{-3/4}) \\
&= o(1).
\end{aligned}
$$

The bound in (8.10) holds since $\lambda! \geq \left(\frac{\lambda}{e}\right)^\lambda$ and $\frac{u_{k+1}}{u_k} > 100$ for $k \leq \lambda$, where

$$
u_k = \frac{(\log n)^k}{k!} \cdot \frac{(\log n)^{(\theta+1)/100} \cdot e^{-\omega/2}}{n \log n}.
$$

□

***Claim* 2** *W.h.p. $\mathbb{G}_{n,p_1}$ does not have a 4-cycle containing a small vertex.*

Proof

$$
\begin{aligned}
&\mathbb{P}(\exists \text{a 4-cycle containing a small vertex}) \\
&\qquad \leq 4n^4 p_1^4 \sum_{k=0}^{(\log n)/100} \binom{n-4}{k} p_1^k (1 - p_1)^{n-4-k} = o(1). \qquad (8.11)
\end{aligned}
$$

□

Claim 3 *Let K be as in Lemma 8.4. Then w.h.p. in $\mathbb{G}_{n,p_1}$, $e(S) < \frac{|S|\log n}{K}$ for every $S \subseteq [n]$, $|S| \le \frac{n}{2eK}$.*

Proof

$$\mathbb{P}\left(\exists |S| \le \frac{n}{2eK} \text{ and } e(S) \ge \frac{|S|\log n}{K}\right) \le \sum_{s=\log n/K}^{n/2eK} \binom{n}{s}\binom{\binom{s}{2}}{s\log n/K} p_1^{s\log n/K}$$
$$= o(1). \tag{8.12}$$

□

Claim 4 *Let K be as in Lemma 8.4. Then, w.h.p. in $\mathbb{G}_{n,p_1}$, if $S \subseteq LARGE$, $|S| \le \frac{n}{2e(\theta+5)K}$, then $|N(S)| \ge (\theta+4)|S|$.*

Proof Let $T = N(S)$, $s = |S|$, $t = |T|$. Then we have

$$e(S \cup T) \ge e(S,T) \ge \frac{|S|\log n}{100} - 2e(S) \ge \frac{|S|\log n}{100} - \frac{2|S|\log n}{K}.$$

Then if $|T| \le (\theta+4)|S|$, we have $|S \cup T| \le (\theta+5)|S| \le \frac{n}{2eK}$ and

$$e(S \cup T) \ge \frac{|S \cup T|}{\theta+5}\left(\frac{1}{100} - \frac{2}{K}\right)\log n = \frac{|S \cup T|\log n}{K}.$$

This contradicts Claim 3.

We can now complete the proof of Lemma 8.4. Let $|S| \le \frac{n}{2e(\theta+5)K}$ and assume that $\mathbb{G}_{n,p_1}$ has minimum degree at least $\theta+1$.

Let $S_1 = S \cap SMALL$ and $S_2 = S \setminus S_1$. Then

$$\begin{aligned} &|N(S)| \\ &\ge |N(S_1)| + |N(S_2)| - |N(S_1) \cap S_2| - |N(S_2) \cap S_1| - |N(S_1) \cap N(S_2)| \\ &\ge |N(S_1)| + |N(S_2)| - |S_2| - |N(S_2) \cap S_1| - |N(S_1) \cap N(S_2)|. \end{aligned}$$

But Claims 1 and 2 and minimum degree at least $\theta+1$ imply that

$$|N(S_1)| \ge (\theta+1)|S_1|,\ |N(S_2) \cap S_1| \le \min\{|S_1|, |S_2|\}, |N(S_1) \cap N(S_2)| \le |S_2|.$$

So, from this and Claim 4 we obtain

$$|N(S)| \ge (\theta+1)|S_1| + (\theta+4)|S_2| - 3|S_2| = (\theta+1)|S|.$$

□

Now go back to the proof of Theorem 8.2 for the case $c = \omega \to \infty$. Let edges of $\mathbb{G}_{n,p_2}$ be $\{f_1, f_2, \ldots, f_s\}$ in random order, where $s \sim \omega n/4$. Let $\mathbb{G}_0 = \mathbb{G}_{n,p_1}$ and $\mathbb{G}_i = \mathbb{G}_{n,p_1} + \{f_1, f_2, \ldots, f_i\}$ for $i \ge 1$. It follows from Lemmas 8.3 and 8.4 that with

$\mu(G)$ as in (8.5), and if $\mu(\mathbb{G}_i) < n/2$ then, assuming that $\mathbb{G}_{n,p_1}$ has the expansion claimed in Lemma 8.4, with $\theta = 0$ and $\alpha = \frac{1}{10eM}$,

$$\mathbb{P}(\mu(\mathbb{G}_{i+1}) \geq \mu(\mathbb{G}_i) + 1 \mid f_1, f_2, \ldots, f_i) \geq \frac{\alpha^2}{2}, \tag{8.13}$$

see (8.9).

It follows that

$$\mathbb{P}(\mathbb{G}_{n,p} \text{ does not have a perfect matching}) \leq o(1) + \mathbb{P}(\text{Bin}(s, \alpha^2/2) < n/2) = o(1).$$

To justify the use of the binomial distribution in the last inequality we used the notion of stochastic dominance and Lemma 2.29. □

Exercises

8.1.1 Verify equation (8.4).
8.1.2 Verify equation (8.11).
8.1.3 Verify equation (8.12).

8.2 Long Paths and Cycles

In this section we study the length of the longest path and cycle in $\mathbb{G}_{n,p}$ when $p = c/n$, where $c = O(\log n)$, most importantly for c is a large constant. We have seen in Section 3.2 that under these conditions, $\mathbb{G}_{n,p}$ will w.h.p. have isolated vertices and so it will not be Hamiltonian. We can however show that it contains a cycle of length $\Omega(n)$ w.h.p.

The question of the existence of a long path/cycle was posed by Erdős and Rényi in [43]. The first positive answer to this question was given by Ajtai, Komlós, and Szemerédi [3] and by de la Vega [111]. The proof we give here is due to Krivelevich, Lee, and Sudakov. It is subsumed by the more general results of [76].

Theorem 8.5 *Let $p = c/n$, where c is sufficiently large but $c = O(\log n)$. Then w.h.p.*

(a) $\mathbb{G}_{n,p}$ has a path of length at least $\left(1 - \frac{6 \log c}{c}\right) n$,
(b) $\mathbb{G}_{n,p}$ has a cycle of length at least $\left(1 - \frac{12 \log c}{c}\right) n$.

Proof We prove this theorem by analyzing simple properties of depth-first search (DFS). This is a well-known algorithm for exploring the vertices of a component of a graph. We can describe the progress of this algorithm using three sets: U is the set of *unexplored* vertices that have not yet been reached by the search. D is the set of *dead* vertices. These have been fully explored and no longer take part in the process. $A = \{a_1, a_2, \ldots, a_r\}$ is the set of active vertices, and they form a path from a_1 to a_r. We start the algorithm by choosing a vertex v from which to start the process. Then we let

$$A = \{v\} \text{ and } D = \emptyset \text{ and } U = [n] \setminus \{v\} \text{ and } r = 1.$$

We now describe how these sets change during one step of the algorithm DFS.

Step (a) If there is an edge $\{a_r, w\}$ for some $w \in U$, then we choose one such w and extend the path defined by A to include w:

$$a_{r+1} \leftarrow w; A \leftarrow A \cup \{w\}; U \leftarrow U \setminus \{w\}; r \leftarrow r+1.$$

We now repeat Step (a).

If there is no such w, then we perform Step (b).

Step (b) We have now completely explored a_r.

$$D \leftarrow D \cup \{a_r\}; A \leftarrow A \setminus \{a_r\}; r \leftarrow r-1.$$

If $r \geq 1$, we go to Step (a). Otherwise, if $U = \emptyset$ at this point, then we terminate the algorithm. If $U \neq \emptyset$, then we choose some $v \in U$ to restart the process with $r = 1$. We then go to Step (a).

We make the following simple observations:

- A step of the algorithm increases $|D|$ by one or decreases $|U|$ by one, and so at some stage we must have $|D| = |U| = s$ for some positive integer s.
- There are no edges between D and U because we only add a_r to D when there are no edges from a_r to U and U does not increase from this point on.

Thus at some stage we have two disjoint sets D, U of size s with no edges between them and a path of length $|A| - 1 = n - 2s - 1$. This plus the following claim implies that $\mathbb{G}_{n,p}$ has a path P of length at least $\left(1 - \frac{6 \log c}{c}\right) n$ w.h.p. Note that if c is large then

$$\alpha > \frac{3 \log c}{c} \text{ implies } c > \frac{2}{\alpha} \log\left(\frac{e}{\alpha}\right).$$

Claim 5 *Let $0 < \alpha < 1$ be a positive constant. If $p = c/n$ and $c > \frac{2}{\alpha} \log\left(\frac{e}{\alpha}\right)$, then w.h.p. in $\mathbb{G}_{n,p}$, every pair of disjoint sets S_1, S_2 of size at least $\alpha n - 1$ are joined by at least one edge.*

Proof The probability that there exist sets S_1, S_2 of size (at least) $\alpha n - 1$ with no joining edge is at most

$$\binom{n}{\alpha n - 1}^2 (1-p)^{(\alpha n - 1)^2} = o(1). \tag{8.14}$$

□

To complete the proof of the theorem, we apply Claim 5 to the vertices S_1, S_2 on the two subpaths P_1, P_2 of length $\frac{3 \log c}{c} n$ at each end of P. There will w.h.p. be an edge joining S_1, S_2, creating the cycle of the claimed length. □

Krivelevich and Sudakov [77] used DFS to give simple proofs of good bounds on the size of the largest component in $\mathbb{G}_{n,p}$ for $p = \frac{1+\varepsilon}{n}$, where ε is a small constant. Problems 8.4.19, 8.4.20, and 8.4.21 elaborate on their results.

Exercises

8.2.1 Verify equation (8.14).

8.3 Hamilton Cycles

This was a difficult question left open in [43]. A breakthrough came with the result of Pósa [101]. The precise theorem given below can be credited to Komlós and Szemerédi [73], Bollobás [20], and Ajtai, Komlós and Szemerédi [4].

Theorem 8.6 *Let* $p = \frac{\log n + \log\log n + c_n}{n}$. *Then*

$$\lim_{n\to\infty} \mathbb{P}(\mathbb{G}_{n,p} \text{ has a Hamilton cycle}) = \begin{cases} 0 & \text{if } c_n \to -\infty, \\ e^{-e^{-c}} & \text{if } c_n \to c, \\ 1 & \text{if } c_n \to \infty. \end{cases}$$

Moreover,

$$\lim_{n\to\infty} \mathbb{P}(\mathbb{G}_{n,p} \text{ has a Hamilton cycle }) = \lim_{n\to\infty} \mathbb{P}(\delta(\mathbb{G}_{n,p}) \geq 2).$$

Proof We will first give a proof of the first statement under the assumption that $c_n = \omega \to \infty$, where $\omega = o(\log\log n)$. Under this assumption, we have $\delta(G_{n,p}) \geq 2$ w.h.p., see Theorem 6.3. The result for larger p follows by monotonicity.

We now set up the main tool, viz. Pósa's lemma. Let P be a path with endpoints a, b, as in Figure 8.1. Suppose that b does not have a neighbor outside of P.

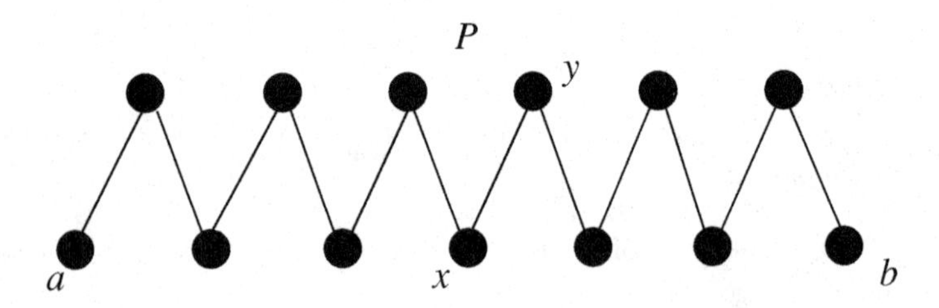

Figure 8.1 The path P

Notice that P' in Figure 8.2 is a path of the same length as P, obtained by a *rotation* with vertex a as the *fixed endpoint*. To be precise, suppose that $P = (a, \ldots, x, y, y', \ldots, b', b)$ and $\{b, x\}$ is an edge where x is an interior vertex of P. The path $P' = (a, \ldots, x, b, b', \ldots, y', y)$ is said to be obtained from P by a rotation.

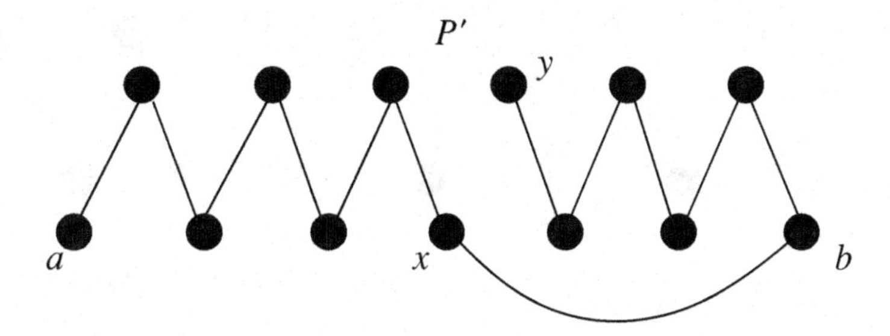

Figure 8.2 The path P' obtained after a single rotation

Now let $END = END(P)$ denote the set of vertices v such that there exists a path P_v from a to v such that P_v is obtained from P by a sequence of rotations with vertex a fixed as in Figure 8.3.

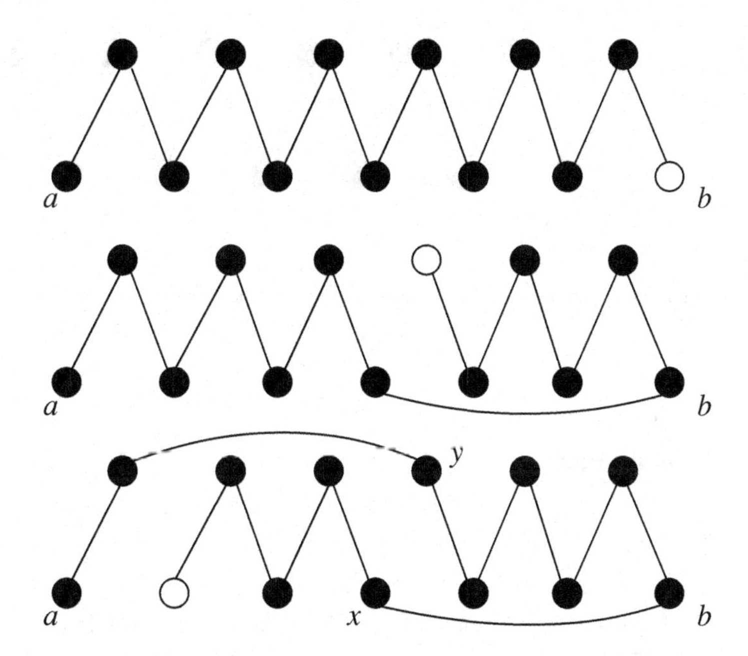

Figure 8.3 A sequence of rotations

Here the set END consists of all the white vertices on the path drawn in Figure 8.4.

Lemma 8.7 *If $v \in P \setminus END$ and v is adjacent to $w \in END$, then there exists $x \in END$ such that the edge $\{v,x\} \in P$.*

Proof Suppose to the contrary that x, y are the neighbors of v on P, that $v, x, y \notin END$ and that v is adjacent to $w \in END$. Consider the path P_w. Let $\{r,t\}$ be the

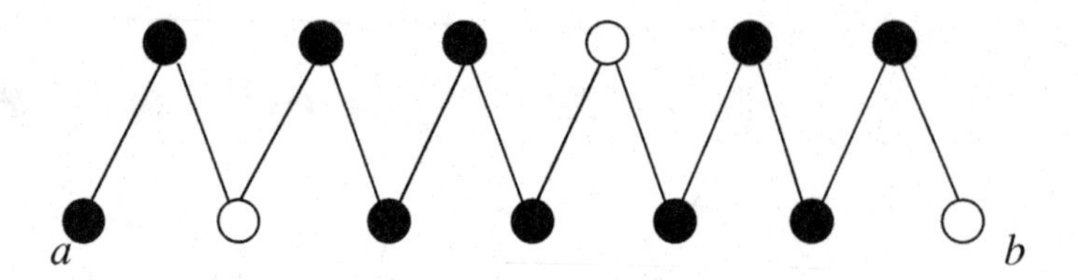

Figure 8.4 The set END

neighbors of v on P_w. Now $\{r,t\} = \{x,y\}$ because if a rotation deletes $\{v,y\}$ say, then v or y becomes an endpoint. But then after a further rotation from P_w we see that $x \in END$ or $y \in END$.

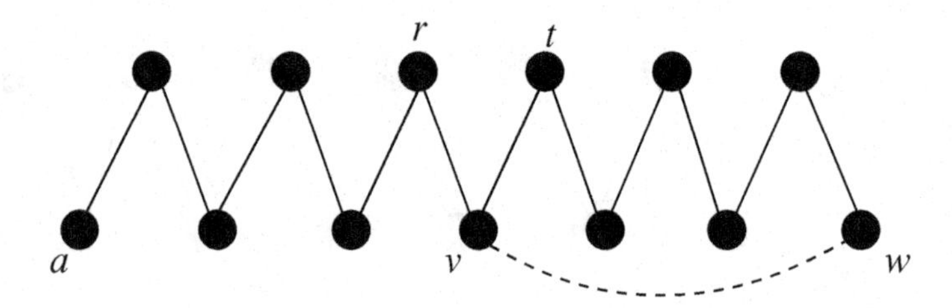

Figure 8.5 One of r, t will become an endpoint after a rotation

□

Corollary 8.8

$$|N(END)| < 2|END|.$$

□

It follows from Lemma 8.4 with $\theta = 1$ that w.h.p. we have

$$|END| \geq \alpha n, \quad \text{where } \alpha = \frac{1}{12eK}. \tag{8.15}$$

We now consider the following algorithm that searches for a Hamilton cycle in a connected graph G. The probability p_1 (as defined in (8.6)) is above the connectivity threshold, and so $\mathbb{G}_{n,p_1}$ is connected w.h.p. Our algorithm will proceed in stages. At the beginning of Stage k we will have a path of length k in G and we will try to grow it by one vertex in order to reach Stage $k+1$. In Stage $n-1$, our aim is simply to create a Hamilton cycle, given a Hamilton path. We start the whole procedure with an arbitrary path of G.

Algorithm Pósa:

(a) Let P be our path at the beginning of Stage k. Let its endpoints be x_0, y_0. If x_0 or y_0 have neighbors outside P, then we can simply extend P to include one of these neighbors and move to Stage $k+1$.

(b) Failing this, we do a sequence of rotations with x_0 as the fixed vertex until one of two things happens: **(i)** We produce a path Q with an endpoint y that has a neighbor outside of Q. In this case we extend Q and proceed to Stage $k+1$. **(ii)** No sequence of rotations leads to Case **(i)**. In this case let END denote the set of endpoints of the paths produced. If $y \in END$, then P_y denotes a path with endpoints x_0, y that is obtained from P by a sequence of rotations.

(c) If we are in Case **(b)(ii)**, then for each $y \in END$ we let $END(y)$ denote the set of vertices z such that there exists a longest path Q_z from y to z such that Q_z is obtained from P_y by a sequence of rotations with vertex y fixed. Repeating the argument above in **(b)** for each $y \in END$, we either extend a path and begin Stage $k+1$ or we go to **(d)**.

(d) Suppose now that we do not reach Stage $k+1$ by an extension and that we have constructed the sets END and $END(y)$ for all $y \in END$. Suppose that G contains an edge (y, z) where $z \in END(y)$. Such an edge would imply the existence of a cycle $C = (z, Q_y, z)$. If this is not a Hamilton cycle, then connectivity implies that there exist $u \in C$ and $v \notin C$ such that u, v are joined by an edge. Let w be a neighbor of u on C and let P' be the path obtained from C by deleting the edge (u, w). This creates a path of length $k+1$, viz. the path w, P', v, and we can move to Stage $k+1$.

A pair z, y where $z \in END(y)$ is called a *booster* in the sense that if we added this edge to $\mathbb{G}_{n,p_1}$ then it would either make the graph Hamiltonian or make the current path longer. We argue now argue that $\mathbb{G}_{n,p_2}$ can be used to "boost" P to a Hamilton cycle, if necessary. We observe now that when $G = \mathbb{G}_{n,p_1}$, $|END| \geq \alpha n$ w.h.p., see (8.15). Also, $|END(y)| \geq \alpha n$ for all $y \in END$. So we will have $\Omega(n^2)$ boosters.

For a graph G let $\lambda(G)$ denote the length of a longest path in G when G is not Hamiltonian and let $\lambda(G) = n$ when G is Hamiltonian. Let the edges of $\mathbb{G}_{n,p_2}$ be $\{f_1, f_2, \ldots, f_s\}$ in random order, where $s \sim \omega n/4$. Let $\mathbb{G}_0 = \mathbb{G}_{n,p_1}$ and $\mathbb{G}_i = \mathbb{G}_{n,p_1} + \{f_1, f_2, \ldots, f_i\}$ for $i \geq 1$. It follows from Lemmas 8.3 and 8.4 that if $\lambda(\mathbb{G}_i) < n$, then, assuming that $\mathbb{G}_{n,p_1}$ has the expansion claimed in Lemma 8.4,

$$\mathbb{P}(\lambda(\mathbb{G}_{i+1}) \geq \lambda(\mathbb{G}_i) + 1 \mid f_1, f_2, \ldots, f_i) \geq \frac{\alpha^2}{2}, \tag{8.16}$$

see (8.9), replacing $A(v)$ by $END(v)$.

It follows that

$$\mathbb{P}(\mathbb{G}_{n,p} \text{ is not Hamiltonian}) \leq o(1) + \mathbb{P}(\text{Bin}(s, \alpha^2/2) < n) = o(1). \tag{8.17}$$

To complete the proof we need to discuss the case where $c_n \to c$. So choose $p_2 = \frac{1}{n \log\log n}$ and let

$$1 - p = (1 - p_1)(1 - p_2).$$

Then we apply Theorem 8.5(a) to argue that w.h.p. G_{n,p_1} has a path of length $n\left(1 - O\left(\frac{\log\log n}{\log n}\right)\right)$.

Now, conditional on G_{n,p_1} having minimum degree at least 2, the proof of the statement of Lemma 8.4 goes through without change for $\theta = 1$, i.e., $S \subseteq [n], |S| \leq \frac{n}{10{,}000}$ implies $|N(S)| \geq 2|S|$. We can then use the extension-rotation argument that we used to prove Theorem 8.6 in the case when $c_n \to \infty$. This time we only need to close $O\left(\frac{n \log\log n}{\log n}\right)$ cycles and we have $\Omega\left(\frac{n}{\log\log n}\right)$ edges. Thus (8.17) is replaced by

$$\mathbb{P}(\mathbb{G}_{n,p} \text{ is not Hamiltonian} \mid \delta(G_{n,p_1}) \geq 2)$$
$$\leq o(1) + \mathbb{P}\left(\text{Bin}\left(\frac{c_1 n}{\log\log n}, 10^{-8}\right) < \frac{c_2 n \log\log n}{\log n}\right) = o(1) \quad (8.18)$$

for some constants c_1, c_2. We used Corollary 2.26 for the second inequality.

Exercises

8.3.1 Verify equation (8.18).

8.4 Spanning Subgraphs

Consider a fixed sequence $H^{(d)}$ of graphs where $n = |V(H^{(d)})| \to \infty$. In particular, we consider a sequence Q_d of d-dimensional cubes where $n = 2^d$ and a sequence of 2-dimensional lattices L_d of order $n = d^2$. We ask when $\mathbb{G}_{n,p}$ or $\mathbb{G}_{n,m}$ contains a copy of $H = H^{(d)}$ w.h.p. We give a condition that can be proved in quite an elegant and easy way. This proof is from Alon and Füredi [8].

Theorem 8.9 *Let H be fixed sequence of graphs with $n = |V(H)| \to \infty$ and maximum degree Δ, where $(\Delta^2 + 1)^2 < n$. If*

$$p^\Delta > \frac{10 \log\lfloor n/(\Delta^2+1)\rfloor}{\lfloor n/(\Delta^2+1)\rfloor}, \quad (8.19)$$

then $\mathbb{G}_{n,p}$ contains an isomorphic copy of H w.h.p.

Proof To prove this we first apply the Hajnal–Szemerédi theorem to the square H^2 of our graph H. Recall that we square a graph if we add an edge between any two vertices of our original graph which are at distance 2. The Hajnal–Szemerédi theorem states that every graph with n vertices and maximum vertex degree at most d is $d+1$-colorable with all color classes of size $\lfloor n/(d+1)\rfloor$ or $\lceil n/(d+1)\rceil$, i.e, the $(d+1)$-coloring is equitable.

Since the maximum degree of H^2 is at most Δ^2, there exists an equitable $\Delta^2 + 1$-coloring of H^2 which induces a partition of the vertex set of H, say $U = U(H)$, into $\Delta^2 + 1$ pairwise disjoint subsets $U_1, U_2, \ldots, U_{\Delta^2+1}$, so that each U_k is an independent set in H^2 and the cardinality of each subset is either $\lfloor n/(\Delta^2 + 1)\rfloor$ or $\lceil n/(\Delta^2 + 1)\rceil$.

Next, partition the set V of vertices of the random graph $\mathbb{G}_{n,p}$ into pairwise disjoint sets $V_1, V_2, \ldots, V_{\Delta^2+1}$, so that $|U_k| = |V_k|$ for $k = 1, 2, \ldots, \Delta^2 + 1$.

We define a one-to-one function $f : U \mapsto V$, which maps each U_k onto V_k, resulting in a mapping of H into an isomorphic copy of H in $\mathbb{G}_{n,p}$. In the first step, choose an arbitrary mapping of U_1 onto V_1. Now U_1 is an independent subset of H and so $\mathbb{G}_{n,p}[V_1]$ trivially contains a copy of $H[U_1]$. Assume, by induction, that we have already defined

$$f : U_1 \cup U_2 \cup \cdots \cup U_k \mapsto V_1 \cup V_2 \cup \cdots \cup V_k,$$

and that f maps the induced subgraph of H on $U_1 \cup U_2 \cup \cdots \cup U_k$ into a copy of it in $V_1 \cup V_2 \cup \cdots \cup V_k$. Now, define f on U_{k+1} using the following construction. Suppose first that $U_{k+1} = \{u_1, u_2, \ldots, u_m\}$ and $V_{k+1} = \{v_1, v_2, \ldots, v_m\}$, where $m \in \{\lfloor n/(\Delta^2 + 1)\rfloor, \lceil n/(\Delta^2 + 1)\rceil\}$.

Next, construct a random bipartite graph $G^{(k)}_{m,m,p^*}$ with a vertex set $V = (X, Y)$, where $X = \{x_1, x_2, \ldots, x_m\}$ and $Y = \{y_1, y_2, \ldots, y_m\}$, and connect x_i and y_j with an edge if and only if in $\mathbb{G}_{n,p}$ the vertex v_j is joined by an edge to all vertices $f(u)$, where u is a neighbor of u_i in H which belongs to $U_1 \cup U_2 \cup \cdots \cup U_k$. Hence, we join x_i with y_j if and only if we can define $f(u_i) = v_j$.

Note that for each i and j, the edge probability $p^* \geq p^\Delta$ and that edges of $G^{(k)}_{m,m,p^*}$ are independent of each other since they depend on pairwise disjoint sets of edges of $\mathbb{G}_{n,p}$. This follows from the fact that U_{k+1} is independent in H^2. Assuming that condition (8.19) holds and that $(\Delta^2 + 1)^2 < n$, then by Theorem 8.1, the random graph $G^{(k)}_{m,m,p^*}$ has a perfect matching w.h.p. Moreover, we can conclude that the probability that there is no perfect matching in $G^{(k)}_{m,m,p^*}$ is at most $\frac{1}{(\Delta^2+1)n}$. It is here that we have used the extra factor 10 on the RHS of (8.19). We use a perfect matching in $G^{(k)}(m, m, p^*)$ to define f, assuming that if x_i and y_j are matched then $f(u_i) = v_j$. To define our mapping $f : U \mapsto V$ we have to find perfect matchings in all $G^{(k)}(m, m, p^*)$, $k = 1, 2, \ldots, \Delta^2+1$. The probability that we can succeed in this is at least $1 - 1/n$. This implies that $\mathbb{G}_{n,p}$ contains an isomorphic copy of H w.h.p. □

Exercises

8.4.1 Let $n = 2^d$ and suppose that $d \to \infty$ and $p \geq \frac{1}{2} + o_d(1)$, where $o_d(1)$ is a function that tends to zero as $d \to \infty$. Show that w.h.p. $\mathbb{G}_{n,p}$ contains a copy of a d-dimensional cube Q_d.

8.4.2 Let $n = d^2$ and $p = \left(\frac{\omega(n)\log n}{n}\right)^{1/4}$, where $\omega(n), d \to \infty$. Show that w.h.p. $\mathbb{G}_{n,p}$ contains a copy of the 2-dimensional lattice L_d.

Problems for Chapter 8

8.1 Consider the bipartite graph process $\Gamma_m, m = 0, 1, 2, \ldots, n^2$, where we add the n^2 edges in $A \times B$ in random order, one by one. Show that w.h.p. the hitting time for Γ_m to have a perfect matching is identical with the hitting time for minimum degree at least 1.

8.2 Show that

$$\lim_{\substack{n\to\infty \\ n \text{ odd}}} \mathbb{P}(\mathbb{G}_{n,p} \text{ has a near perfect matching}) = \begin{cases} 0 & \text{if } c_n \to -\infty, \\ e^{-c-e^{-c}} & \text{if } c_n \to c, \\ 1 & \text{if } c_n \to \infty. \end{cases}$$

A *near perfect matching* is one of size $\lfloor n/2 \rfloor$.

8.3 Show that if $p = \frac{\log n+(k-1)\log\log n+\omega}{n}$, where $k = O(1)$ and $\omega \to \infty$, then w.h.p. $G_{n,n,p}$ contains a k-regular spanning subgraph.

8.4 Consider the random bipartite graph G with bipartition A, B, where $|A| = |B| = n$. Each vertex $a \in A$ independently chooses $\lceil 2\log n \rceil$ random neighbors in B. Show that w.h.p. G contains a perfect matching.

8.5 Let $G = (X, Y, E)$ be an arbitrary bipartite graph where the bipartition X, Y satisfies $|X| = |Y| = n$. Suppose that G has minimum degree at least $3n/4$. Let $p = \frac{K\log n}{n}$, where K is a large constant. Show that w.h.p. G_p contains a perfect matching.

8.6 Suppose that $n = 2m$ is even, that $np \gg \log n$ and that ε is an arbitrary positive constant. Show that w.h.p. $G_{n,p} \setminus H$ contains a perfect matching for all choices of subgraph H whose maximum degree $\Delta(H) \le \left(\frac{1}{2} - \varepsilon\right) np$.

8.7 Show that if $p = \frac{\log n+(k-1)\log\log n+\omega}{n}$, where $k = O(1)$ and $\omega \to \infty$, then w.h.p. $G_{n,p}$ contains $\lfloor k/2 \rfloor$ edge disjoint Hamilton cycles. If k is odd, show that in addition there is an edge disjoint matching of size $\lfloor n/2 \rfloor$. (Hint: Use Lemma 8.4 to argue that after "peeling off" a few Hamilton cycles, we can still use the arguments of Sections 8.1 and 8.3.)

8.8 Let m_k^* denote the first time that G_m has minimum degree at least k. Show that w.h.p. in the graph process (i) $G_{m_1^*}$ contains a perfect matching and (ii) $G_{m_2^*}$ contains a Hamilton cycle.

8.9 Show that if $p = \frac{\log n+\log\log n+\omega}{n}$, where $\omega \to \infty$, then w.h.p. $G_{n,n,p}$ contains a Hamilton cycle. (Hint: Start with a 2-regular spanning subgraph from Problem 8.3. Delete an edge from a cycle. Argue that rotations will always produce paths beginning and ending at different sides of the partition. Proceed more or less as in Section 8.3.)

8.10 Show that if $p = \frac{\log n+\log\log n+\omega}{n}$, where n is even and $\omega \to \infty$, then w.h.p. $G_{n,p}$ contains a pair of vertex disjoint $n/2$-cycles. (Hint: Randomly partition $[n]$ into two sets of size $n/2$. Then move some vertices between parts to make the minimum degree at least 2 in both parts.)

8.11 Show that if 3 divides n and $np^2 \gg \log n$, then w.h.p. $G_{n,p}$ contains $n/3$ vertex disjoint triangles. (Hint: Randomly partition $[n]$ into three sets A, B, C of size $n/3$. Choose a perfect matching M between A and B and then match C into M.)

8.12 Let $p = (1+\varepsilon)\frac{\log n}{n}$ for some fixed $\varepsilon > 0$. Prove that w.h.p. $G_{n,p}$ is Hamilton connected, i.e., every pair of vertices are the endpoints of a Hamilton path.

8.13 Show that if $p = \frac{(1+\varepsilon)\log n}{n}$ for $\varepsilon > 0$ constant, then w.h.p. $G_{n,p}$ contains a copy of a caterpillar on n vertices. The diagram below is the case $n = 16$.

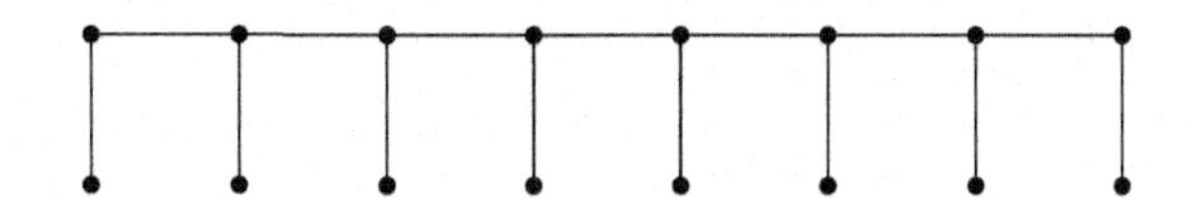

8.14 Show that for any fixed $\varepsilon > 0$ there exists c_ε such that if $c \geq c_\varepsilon$ then $\mathbb{G}_{n,p}$ contains a cycle of length $(1-\varepsilon)n$ with probability $1 - e^{-c\varepsilon^2 n/10}$.

8.15 Let $p = (1+\varepsilon)\frac{\log n}{n}$ for some fixed $\varepsilon > 0$. Prove that w.h.p. $G_{n,p}$ is pancyclic, i.e., it contains a cycle of length k for every $3 \leq k \leq n$. (See Cooper and Frieze [36] and Cooper [34, 35].)

8.16 Show that if p is constant then

$$\mathbb{P}(\mathbb{G}_{n,p} \text{ is not Hamiltonian}) = O(e^{-\Omega(np)}).$$

8.17 Let T be a tree on n vertices and maximum degree less than $c_1 \log n$. Suppose that T has at least $c_2 n$ leaves. Show that there exists $K = K(c_1, c_2)$ such that if $p \geq \frac{K \log n}{n}$ then $G_{n,p}$ contains a copy of T w.h.p.

8.18 Let $p = \frac{1000}{n}$ and $G = G_{n,p}$. Show that w.h.p. any red-blue coloring of the edges of G contains a monochromatic path of length $\frac{n}{1000}$. (Hint: Apply the argument of Section 8.2 to both the red and blue subgraphs of G to show that if there is no long monochromatic path, then there is a pair of large sets S, T such that no edge joins S, T. This question is taken from Dudek and Prałat [40].

8.19 Suppose that $p = n^{-\alpha}$ for some constant $\alpha > 0$. Show that if $\alpha > \frac{1}{3}$, then w.h.p. $\mathbb{G}_{n,p}$ does not contain a maximal spanning planar subgraph, i.e., a planar subgraph with $3n - 6$ edges. Show that if $\alpha < \frac{1}{3}$ then it contains one w.h.p. (see Bollobás and Frieze [26]).

8.20 Show that the hitting time for the existence of k edge-disjoint spanning trees coincides w.h.p. with the hitting time for minimum degree k for $k = O(1)$ (see Palmer and Spencer [97]).

8.21 Consider the modified greedy matching algorithm where you first choose a random vertex x and then choose a random edge $\{x, y\}$ incident with x. Show that applied to $\mathbb{G}_{n,m}$, with $m = cn$, that w.h.p. it produces a matching of size $\left(\frac{1}{2} + o(1) - \frac{\log(2-e^{-2c})}{4c}\right) n$.

8.22 Let $X_1, X_2, \ldots, X_N$, $N = \binom{n}{2}$ be a sequence of independent Bernouilli random variables with common probability p. Let $\varepsilon > 0$ be sufficiently small.

(a) Let $p = \frac{1-\varepsilon}{n}$ and let $k = \frac{7 \log n}{\varepsilon^2}$. Show that w.h.p. there is no interval I of length kn in $[N]$ in which at least k of the variables take the value 1.

(b) Let $p = \frac{1+\varepsilon}{n}$ and let $N_0 = \frac{\varepsilon n^2}{2}$. Show that w.h.p.

$$\left| \sum_{i=1}^{N_0} X_i - \frac{\varepsilon(1+\varepsilon)n}{2} \right| \leq n^{2/3}.$$

8.23 Use the result of Problem 8.22(a) to show that if $p = \frac{1-\varepsilon}{n}$ then w.h.p. the maximum component size in $\mathbb{G}_{n,p}$ is at most $\frac{7 \log n}{\varepsilon^2}$. (See [77].)

8.24 Use the result of Problem 8.22(b) to show that if $p = \frac{1+\varepsilon}{n}$ then w.h.p $\mathbb{G}_{n,p}$ contains a path of length at least $\frac{\varepsilon^2 n}{5}$.

9 Extreme Characteristics

This chapter is devoted to the extremes of certain graph parameters. The extreme values of various statistics form the basis of a large part of graph theory. From Turan's theorem to Ramsey theory, there has been an enormous amount of research on which graph parameters affect these statistics. One of the most important results is that of Erdős, who showed that the absence of small cycles did not guarantee a bound on the chromatic number. This incidentally was one of the first uses of random graphs to prove deterministic results.

In this chapter, we look first at the diameter of random graphs, i.e., the extreme value of the shortest distance between a pair of vertices. Then we look at the size of the largest independent set and the related value of the chromatic number. One interesting feature of these parameters is that they are often highly concentrated. In some cases one can say that w.h.p. they will have one of two possible values.

9.1 Diameter

In this section we will first discuss the threshold for $\mathbb{G}_{n,p}$ to have diameter d when $d \geq 2$ is a constant. The diameter $\text{diam}(G)$ of a connected graph G is the maximum over distinct vertices v, w of $\text{dist}(v, w)$ where $\text{dist}(v, w)$ is the minimum number of edges in a path from v to w.

Theorem 9.1 *Let $d \geq 2$ be a fixed positive integer. Suppose that $c > 0$ and*

$$p^d n^{d-1} = \log(n^2/c).$$

Then w.h.p. $\text{diam}(\mathbb{G}_{n,p})$ *is either d or $d + 1$.*

Proof (a) w.h.p. $\text{diam}(G) \geq d$.
Fix $v \in V$ and let

$$N_k(v) = \{w : \text{dist}(v, w) = k\}. \tag{9.1}$$

It follows from Theorem 5.5 that w.h.p. for $0 \leq k < d$,

$$|N_k(v)| \leq \Delta^k \sim (np)^k \sim (n \log n)^{k/d} = o(n). \tag{9.2}$$

(b) w.h.p. $\text{diam}(G) \leq d + 1$.

Fix $v, w \in [n]$. Then for $1 \leq k < d$, define the event

$$\mathcal{F}_k = \left\{|N_k(v)| \in I_k = \left[\left(\frac{np}{2}\right)^k, (2np)^k\right]\right\}.$$

Then for $k \leq \lceil d/2 \rceil$ we have

$$\begin{aligned}
&\mathbb{P}(\text{not } \mathcal{F}_k \mid \mathcal{F}_1, \ldots, \mathcal{F}_{k-1}) \\
&= \mathbb{P}\left(\text{Bin}\left(n - \sum_{i=0}^{k-1} |N_i(v)|, 1 - (1-p)^{|N_{k-1}(v)|}\right) \notin I_k\right) \\
&\leq \mathbb{P}\left(\text{Bin}\left(n - o(n), \frac{3}{4}\left(\frac{np}{2}\right)^{k-1} p\right) \leq \left(\frac{np}{2}\right)^k\right) \\
&\quad + \mathbb{P}\left(\text{Bin}\left(n - o(n), \frac{5}{4}(2np)^{k-1} p\right) \geq (2np)^k\right) \\
&\leq \exp\left\{-\Omega\left((np)^k\right)\right\} \\
&= O(n^{-3}).
\end{aligned}$$

So with probability $1 - O(n^{-3})$,

$$|N_{\lfloor d/2 \rfloor}(v)| \geq \left(\frac{np}{2}\right)^{\lfloor d/2 \rfloor} \quad \text{and} \quad |N_{\lceil d/2 \rceil}(w)| \geq \left(\frac{np}{2}\right)^{\lceil d/2 \rceil}.$$

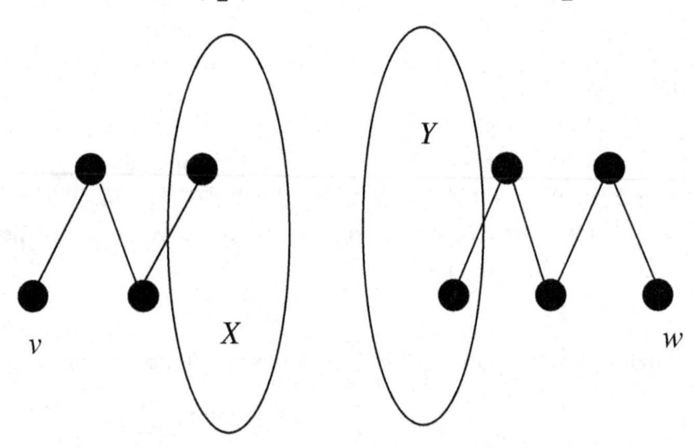

If $X = N_{\lfloor d/2 \rfloor}(v)$ and $Y = N_{\lceil d/2 \rceil}(w)$, then either

$$X \cap Y \neq \emptyset \quad \text{and} \quad \text{dist}(v, w) \leq \lfloor d/2 \rfloor + \lceil d/2 \rceil = d,$$

or since the edges between X and Y are unconditioned by our construction,

$$\begin{aligned}
\mathbb{P}(\nexists \text{ an } X : Y \text{ edge }) &\leq (1-p)^{\left(\frac{np}{2}\right)^d} \leq \exp\left\{-\left(\frac{np}{2}\right)^d p\right\} \\
&\leq \exp\{-(2 - o(1))np \log n\} = o(n^{-3}).
\end{aligned}$$

So

$$\mathbb{P}(\exists v, w : \text{dist}(v, w) > d + 1) = o(n^{-1}).$$

□

This theorem can be strengthened to provide the limiting probability that the diameter is actually d or $d + 1$. See for example Theorem 7.1 of [52].

We turn next to a sparser case and prove a somewhat weaker result.

Theorem 9.2 *Suppose that* $p = \frac{\omega \log n}{n}$ *where* $\omega \to \infty$. *Then*

$$\text{diam}(\mathbb{G}_{n,p}) \sim \frac{\log n}{\log np} \qquad w.h.p.$$

Proof Fix $v \in [n]$ and let $N_i = N_i(v)$ be as in (9.1). Let $N_{\leq k} = \bigcup_{i \leq k} N_i$. Using the proof of Theorem 5.5(ii) we see that we can assume that

$$(1 - \omega^{-1/3})np \leq \deg(x) \leq (1 + \omega^{-1/3})np \qquad \text{for all } x \in [n]. \tag{9.3}$$

It follows that if $\gamma = \omega^{-1/3}$ and

$$k_0 = \frac{\log n - \log 3}{\log np + \gamma} \sim \frac{\log n}{\log np},$$

then w.h.p.

$$|N_{\leq k_0}| \leq \sum_{k \leq k_0} ((1 + \gamma)np)^k \leq 2((1 + \gamma)np)^{k_0} = \frac{2n}{3 + o(1)},$$

and so the diameter of $\mathbb{G}_{n,p}$ is at least $(1 - o(1))\frac{\log n}{\log np}$.

We can assume that $np = n^{o(1)}$ as larger p are dealt with in Theorem 9.1. Now fix $v, w \in [n]$ and let N_i be as in the previous paragraph. Now consider a Breadth-First Search (BFS) from v that constructs $N_1, N_2, \ldots, N_{k_1}$ where

$$k_1 = \frac{3 \log n}{5 \log np}.$$

It follows that if (9.3) holds then for $k \leq k_1$ we have

$$|N_{i \leq k}| \leq n^{3/4} \text{ and } |N_k| p \leq n^{-1/5}. \tag{9.4}$$

Observe now that the edges from N_i to $[n] \setminus N_{\leq i}$ are unconditioned by the BFS up to layer k and so for $x \in [n] \setminus N_{\leq k}$,

$$\mathbb{P}(x \in N_{k+1} \mid N_{\leq k}) = 1 - (1 - p)^{|N_k|} \geq \rho_k = |N_k| p(1 - n^{-1/5}). \tag{9.5}$$

The events $x \in N_{k+1}$ are independent, and so $|N_{k+1}|$ stochastically dominates the binomial $\text{Bin}(n - n^{3/4}, \rho_k)$. Assume inductively that $|N_k| \geq (1 - \gamma)^k (np)^k$ for some $k \geq 1$. This is true w.h.p. for $k = 1$ by (9.3). Let $\mathcal{A}_k$ be the event that (9.4) holds. It follows that

$$\mathbb{E}(|N_{k+1}| \mid \mathcal{A}_k) \geq np|N_k|(1 - O(n^{-1/5})).$$

It then follows from the Chernoff–Hoeffding bounds (Theorem 2.20) that

$$\mathbb{P}(|N_{k+1}| \leq ((1-\gamma)np)^{k+1} \leq \exp\left\{-\frac{\gamma^2}{4}|N_k|np\right\} = o(n^{-any\ constant}).$$

There is a small point to be made about conditioning here. We can condition on (9.3) holding and then argue that this only multiplies small probabilities by $1 + o(1)$ if we use $\mathbb{P}(A \mid B) \leq \mathbb{P}(A)/\mathbb{P}(B)$.

It follows that if

$$k_2 = \frac{\log n}{2(\log np + \log(1-\gamma))} \sim \frac{\log n}{2\log np},$$

then w.h.p. we have

$$|N_{k_2}| \geq n^{1/2}.$$

Analogously, if we do BFS from w to create $N'_k, i = 1, 2, \dots, k_2$, then $|N'_{k_2}| \geq n^{1/2}$. If $N_{\leq k_2} \cap N'_{\leq k_2} \neq \emptyset$, then $\text{dist}(v, w) \leq 2k_2$ and we are done. Otherwise, we observe that the edges $E(N_{k_2}, N'_{k_2})$ between N_{k_2} and N'_{k_2} are unconditioned (except for (9.3)) and so

$$\mathbb{P}(E(N_{k_2}, N'_{k_2}) = \emptyset) \leq (1-p)^{n^{1/2} \times n^{1/2}} \leq n^{-\omega}.$$

If $E(N_{k_2}, N'_{k_2}) \neq \emptyset$, then $\text{dist}(v, w) \leq 2k_2 + 1$ and we are done. Note that given (9.3), all other unlikely events have probability $O(n^{-anyconstant})$ of occurring, and so we can inflate these latter probabilities by n^2 to account for all choices of v, w. This completes the proof of Theorem 9.2. □

Exercises

9.1.1 Verify equation (9.5).

9.1.2 Suppose that $0 < p < 1$ is constant. Show that w.h.p. $\mathbb{G}_{n,p}$ has diameter 2.

9.2 Largest Independent Sets

Let $\alpha(G)$ denote the size of the largest independent set in a graph G.

The following theorem for a dense random graph was first proved by Matula [85].

Theorem 9.3 *Suppose* $0 < p < 1$ *is a constant and* $b = \frac{1}{1-p}$. *Then w.h.p.*

$$\alpha(\mathbb{G}_{n,p}) \sim 2\log_b n.$$

Proof Let X_k be the number of independent sets of order k and let

$$k = \lceil 2\log_b n \rceil.$$

Then,

$$\mathbb{E} X_k = \binom{n}{k}(1-p)^{\binom{k}{2}}$$

$$\le \left(\frac{ne}{k(1-p)^{1/2}} (1-p)^{k/2} \right)^k$$
$$\le \left(\frac{e}{k(1-p)^{1/2}} \right)^k$$
$$= o(1).$$

To complete the proof we shall use Janson's inequality. To introduce this inequality we need the following notation. Fix a family of n subsets $D_i, i \in [n]$. Let R be a random subset of $[N]$, $N = \binom{n}{2}$, such that for $s \in [N]$ we have $0 < \mathbb{P}(s \in R) = q_s < 1$. The elements of R are chosen independently of each other and the sets $D_i, i = 1, 2, \dots, n$. Let $\mathcal{A}_i$ be the event that D_i is a subset of R. Moreover, let I_i be the indicator of the event $\mathcal{A}_i$. Note that I_i and I_j are independent if and only if $D_i \cap D_j = \emptyset$. We let

$$S_n = I_1 + I_2 + \cdots + I_n,$$

and

$$\mu = \mathbb{E} S_n = \sum_{i=1}^{n} \mathbb{E}(I_i).$$

We write $i \sim j$ if $D_i \cap D_j \neq \emptyset$. Then, let

$$\overline{\Delta} = \sum_{\{i,j\}: i \sim j} \mathbb{E}(I_i I_j) = \mu + \Delta, \tag{9.6}$$

where

$$\Delta = \sum_{\substack{\{i,j\}: i \sim j \\ i \neq j}} \mathbb{E}(I_i I_j). \tag{9.7}$$

As before, let $\varphi(x) = (1+x)\log(1+x) - x$. Now, with $S_n, \overline{\Delta}, \varphi$ given above, one can establish the following upper bound on the lower tail of the distribution of S_n.

Lemma 9.4 (Janson's inequality) *For any real t, $0 \le t \le \mu$,*

$$\mathbb{P}(S_n \le \mu - t) \le \exp\left\{ -\frac{\varphi(-t/\mu)\mu^2}{\overline{\Delta}} \right\} \le \exp\left\{ -\frac{t^2}{2\overline{\Delta}} \right\}. \tag{9.8}$$

Let now

$$k = \lfloor 2 \log_b n - 5 \log_b \log n \rfloor.$$

Let

$$\overline{\Delta} = \sum_{\substack{i,j \\ S_i \sim S_j}} \mathbb{P}(S_i, S_j \text{ are independent in } \mathbb{G}_{n,p}),$$

where $S_1, S_2, \ldots, S_{\binom{n}{k}}$ are all the k-subsets of $[n]$ and $S_i \sim S_j$ if and only if $|S_i \cap S_j| \geq 2$.

So, finishing the proof, we see that

$$\mathbb{P}(X_k = 0) \leq \exp\left\{-\frac{(\mathbb{E}\,X_k)^2}{2\overline{\Delta}}\right\}.$$

Here we apply the inequality in the context of X_k being the number of k-cliques in the complement of $G_{n,p}$. The set $[N]$ will be the edges of the complete graph, and the sets D_i will be the edges of the k-cliques. Now

$$\begin{aligned}\frac{\overline{\Delta}}{(\mathbb{E}\,X_k)^2} &= \frac{\binom{n}{k}(1-p)^{\binom{k}{2}}\sum_{j=2}^{k}\binom{n-k}{k-j}\binom{k}{j}(1-p)^{\binom{k}{2}-\binom{j}{2}}}{\left(\binom{n}{k}(1-p)^{\binom{k}{2}}\right)^2} \\ &= \sum_{j=2}^{k}\frac{\binom{n-k}{k-j}\binom{k}{j}}{\binom{n}{k}}(1-p)^{-\binom{j}{2}} \\ &= \sum_{j=2}^{k} u_j.\end{aligned}$$

Notice that for $j \geq 2$,

$$\frac{u_{j+1}}{u_j} = \frac{k-j}{n-2k+j+1}\,\frac{k-j}{j+1}(1-p)^{-j} \leq \left(1 + O\left(\frac{\log_b n}{n}\right)\right)\frac{k^2(1-p)^{-j}}{n(j+1)}.$$

Therefore,

$$\begin{aligned}\frac{u_j}{u_2} &\leq (1+o(1))\left(\frac{k^2}{n}\right)^{j-2}\frac{2(1-p)^{-(j-2)(j+1)/2}}{j!} \\ &\leq (1+o(1))\left(\frac{2k^2e}{nj}(1-p)^{-\frac{j+1}{2}}\right)^{j-2} \leq 1.\end{aligned}$$

So

$$\frac{(\mathbb{E}\,X_k)^2}{\overline{\Delta}} \geq \frac{1}{ku_2} \geq \frac{n^2(1-p)}{k^5}.$$

Therefore, by Janson's inequality,

$$\mathbb{P}(X_k = 0) \leq e^{-\Omega(n^2/(\log n)^5)}. \tag{9.9}$$

□

Matula used the Chebyshev inequality, and so he was not able to prove an exponential bound like (9.9). This will be important when we come to discuss the chromatic number. We now consider the case where $p = d/n$ and d is a large constant, i.e., when a random graph is sparse. Frieze [51] proved

Theorem 9.5 *Let $\varepsilon > 0$ be a fixed constant. Then for $d \geq d(\varepsilon)$ we have that w.h.p.*

$$\left|\alpha(\mathbb{G}_{n,p})) - \frac{2n}{d}(\log d - \log\log d - \log 2 + 1)\right| \leq \frac{\varepsilon n}{d}.$$

In this section we will only prove that if $p = d/n$ and d is sufficiently large, then w.h.p.

$$\left|\alpha(\mathbb{G}_{n,p}) - \frac{2\log d}{d}n\right| \le \frac{\varepsilon \log d}{d}n. \tag{9.10}$$

This will follow from the following. Let X_k be as defined above. Let

$$k_0 = \frac{(2-\varepsilon/8)\log d}{d}n \text{ and } k_1 = \frac{(2+\varepsilon/8)\log d}{d}n.$$

Then,

$$\mathbb{P}\left(\left|\alpha(\mathbb{G}_{n,p}) - \mathbb{E}(\alpha(\mathbb{G}_{n,p}))\right| \ge \frac{\varepsilon \log d}{8d}n\right) \le \exp\left\{-\Omega\left(\frac{(\log d)^2}{d^2}\right)n\right\}. \tag{9.11}$$

$$\mathbb{P}(X_{k_1} > 0) \le \exp\left\{-\Omega\left(\frac{(\log d)^2}{d}\right)n\right\}. \tag{9.12}$$

$$\mathbb{P}(X_{k_0} > 0) \ge \exp\left\{-O\left(\frac{(\log d)^{3/2}}{d^2}\right)n\right\}. \tag{9.13}$$

Let us see how (9.10) follows from these three inequalities. Indeed, (9.11) and (9.13) imply that

$$\mathbb{E}(\alpha(\mathbb{G}_{n,p})) \ge k_0 - \frac{\varepsilon \log d}{8d}n. \tag{9.14}$$

Furthermore, (9.11) and (9.12) imply that

$$\mathbb{E}(\alpha(\mathbb{G}_{n,p})) \le k_1 + \frac{\varepsilon \log d}{8d}n. \tag{9.15}$$

It follows from (9.14) and (9.15) that

$$|k_0 - \mathbb{E}(\alpha(\mathbb{G}_{n,p}))| \le \frac{\varepsilon \log d}{2d}n.$$

We obtain (9.10) by applying (9.11) once more.

Proof of (9.11)**:** This follows directly from the following useful lemma.

Lemma 9.6 (McDiarmid's inequality) *Let $Z = Z(W_1, W_2, \dots, W_N)$ be a random variable that depends on N independent random variables $W_1, W_2, \dots, W_N$. Suppose that*

$$|Z(W_1, \dots, W_i, \dots, W_N) - Z(W_1, \dots, W_i', \dots, W_N)| \le c_i$$

for all $i = 1, 2, \dots, N$ and $W_1, W_2, \dots, W_N, W_i'$. Then for all $t > 0$ we have

$$\mathbb{P}(Z \ge \mathbb{E}Z + t) \le \exp\left\{-\frac{t^2}{2\sum_{i=1}^N c_i^2}\right\},$$

and

$$\mathbb{P}(Z \le \mathbb{E}Z - t) \le \exp\left\{-\frac{t^2}{2\sum_{i=1}^N c_i^2}\right\}.$$

If $Z = \alpha(\mathbb{G}_{n,p})$, then we write $Z = Z(Y_2, Y_3, \dots, Y_n)$, where Y_i is the set of edges between vertex i and vertices $[i-1]$ for $i \geq 2$. $Y_2, Y_3, \dots, Y_n$ are independent and changing a single Y_i can change Z by at most 1. Therefore, for any $t > 0$ we have

$$\mathbb{P}(|Z - \mathbb{E}(Z)| \geq t) \leq \exp\left\{-\frac{t^2}{2n-2}\right\}.$$

Setting $t = \frac{\varepsilon \log d}{8d} n$ yields (9.11).

Proof of (9.12)**:** The First Moment Method gives

$$\mathbb{P}(X_{k_1} > 0) \leq \binom{n}{k_1}\left(1 - \frac{d}{n}\right)^{\binom{k_1}{2}} \leq \exp\left\{-\Omega\left(\frac{(\log d)^2}{d}\right) n\right\}. \tag{9.16}$$

Proof of (9.13)**:** Now, after using the upper bound given in Lemma 2.9, we get

$$\frac{1}{\mathbb{P}(X_{k_0} > 0)} \leq \frac{\mathbb{E}(X_{k_0}^2)}{\mathbb{E}(X_{k_0})^2} = \sum_{j=0}^{k_0} \frac{\binom{n-k_0}{k_0-j}\binom{k_0}{j}}{\binom{n}{k_0}} (1-p)^{-\binom{j}{2}}$$

$$\leq \sum_{j=0}^{k_0} \left(\frac{k_0 e}{j} \cdot \exp\left\{\frac{jd}{2n} + O\left(\frac{jd^2}{n^2}\right)\right\}\right)^j \times \left(\frac{k_0}{n}\right)^j \left(\frac{n-k_0}{n-j}\right)^{k_0-j} \tag{9.17}$$

$$\leq \sum_{j=0}^{k_0} \left(\frac{k_0 e}{j} \cdot \frac{k_0}{n} \cdot \exp\left\{\frac{jd}{2n} + O\left(\frac{jd^2}{n^2}\right)\right\}\right)^j \times \exp\left\{-\frac{(k_0-j)^2}{n-j}\right\}$$

$$\leq_b \sum_{j=0}^{k_0} \left(\frac{k_0 e}{j} \cdot \frac{k_0}{n} \cdot \exp\left\{\frac{jd}{2n} + \frac{2k_0}{n}\right\}\right)^j \times \exp\left\{-\frac{k_0^2}{n}\right\}$$

$$= \sum_{j=0}^{k_0} v_j. \tag{9.18}$$

(The notation $A \leq_b B$ is shorthand for $A = O(B)$ when the latter is considered to be ugly looking.)

We first observe that $(A/x)^x \leq e^{A/e}$ for $A > 0$ implies that

$$\left(\frac{k_0 e}{j} \cdot \frac{k_0}{n}\right)^j \times \exp\left\{-\frac{k_0^2}{n}\right\} \leq 1.$$

So,

$$j \leq j_0 = \frac{(\log d)^{3/4}}{d^{3/2}} n \implies v_j \leq \exp\left\{\frac{j^2 d}{2n} + \frac{2jk_0}{n}\right\} = \exp\left\{O\left(\frac{(\log d)^{3/2}}{d^2}\right) n\right\}. \tag{9.19}$$

Now put

$$j = \frac{\alpha \log d}{d} n, \quad \text{where } \frac{1}{d^{1/2}(\log d)^{1/4}} < \alpha < 2 - \frac{\varepsilon}{4}.$$

Then

$$\begin{aligned} \frac{k_0 e}{j} \cdot \frac{k_0}{n} \cdot \exp\left\{\frac{jd}{2n} + \frac{2k_0}{n}\right\} &\leq \frac{4e \log d}{\alpha d} \cdot \exp\left\{\frac{\alpha \log d}{2} + \frac{4 \log d}{d}\right\} \\ &= \frac{4e \log d}{\alpha d^{1-\alpha/2}} \exp\left\{\frac{4 \log d}{d}\right\} \\ &< 1. \end{aligned} \tag{9.20}$$

To see this note that if $f(\alpha) = \alpha d^{1-\alpha/2}$ then f increases between $d^{-1/2}$ and $2/\log d$ after which it decreases. Then note that

$$\min\left\{f(d^{-1/2}), f(2-\varepsilon)\right\} > 4e \exp\left\{\frac{4 \log d}{d}\right\}.$$

Thus $v_j < 1$ for $j \geq j_0$ and (9.13) follows from (9.19). □

Exercises

9.2.1 Verify equation (9.16).

9.2.2 Verify equation (9.20).

9.2.3 Let $p = d/n$, where d is a positive constant. Let S be the set of vertices of degree at least $\frac{2 \log n}{3 \log \log n}$. Show that w.h.p. S is an independent set.

9.2.4 Let $p = d/n$, where d is a large positive constant. Use the First Moment Method to show that w.h.p.

$$\alpha(G_{n,p}) \leq \frac{2n}{d}(\log d - \log \log d - \log 2 + 1 + \varepsilon)$$

for any positive constant ε.

9.3 Chromatic Number

Let $\chi(G)$ denote the chromatic number of a graph G, i.e., the smallest number of colors with which one can properly color the vertices of G. A coloring is proper if no two adjacent vertices have the same color.

We will first describe the asymptotic behavior of the chromatic number of dense random graphs. The following theorem is a major result, due to Bollobás [22]. The upper bound without the 2 in the denominator follows directly from Theorem 9.3. An intermediate result giving 3/2 instead of 2 was already proved by Matula [86].

Theorem 9.7 *Suppose* $0 < p < 1$ *is a constant and* $b = \frac{1}{1-p}$. *Then w.h.p.*

$$\chi(\mathbb{G}_{n,p}) \sim \frac{n}{2\log_b n}.$$

Proof By Theorem 9.3,

$$\chi(\mathbb{G}_{n,p}) \geq \frac{n}{\alpha(\mathbb{G}_{n,p})} \sim \frac{n}{2\log_b n}.$$

Let $\nu = \frac{n}{(\log_b n)^2}$ and $k_0 = 2\log_b n - 4\log_b \log_b n$. It follows from (9.9) that

$$\begin{aligned} &\mathbb{P}(\exists S : |S| \geq w,\ S \text{ does not contain an independent set of order } \geq k_0) \\ &\quad \leq \binom{n}{\nu} \exp\left\{-\Omega\left(\frac{\nu^2}{(\log n)^5}\right)\right\} \qquad (9.21) \\ &\quad = o(1). \qquad (9.22) \end{aligned}$$

So assume that every set of order at least ν contains an independent set of order at least k_0. We repeatedly choose an independent set of order k_0 among the set of uncolored vertices. Give each vertex in this set a new color. Repeat until the number of uncolored vertices is at most ν. Give each remaining uncolored vertex its own color. The number of colors used is at most

$$\frac{n}{k_0} + \nu \sim \frac{n}{2\log_b n}.$$

□

We shall next show that the chromatic number of a dense random graph is highly concentrated.

Theorem 9.8 *Suppose* $0 < p < 1$ *is a constant. Then*

$$\mathbb{P}(|\chi(\mathbb{G}_{n,p}) - \mathbb{E}\,\chi(\mathbb{G}_{n,p})| \geq t) \leq 2\exp\left\{-\frac{t^2}{2n}\right\}.$$

Proof Write

$$\chi = Z(Y_1, Y_2, \ldots, Y_n), \qquad (9.23)$$

where

$$Y_j = \{(i,j) \in E(\mathbb{G}_{n,p}) : i < j\}.$$

Then

$$|Z(Y_1, Y_2, \ldots, Y_n) - Z(Y_1, Y_2, \ldots, \hat{Y}_i, \ldots, Y_n)| \leq 1$$

and the theorem follows by Lemma 9.6. □

Greedy Coloring Algorithm

We show below that a simple greedy algorithm performs very efficiently. It uses twice as many colors as it "should" in the light of Theorem 9.7. This algorithm is discussed in Bollobás and Erdős [24] and Grimmett and McDiarmid [55]. It starts by greedily choosing an independent set C_1 and at the same time giving its vertices color 1. C_1 is removed and then we greedily choose an independent set C_2 and give its vertices color 2 and so on, until all vertices have been colored.

Algorithm *GREEDY*

- k is the current color.
- A is the current set of vertices that might get color k in the current round.
- U is the current set of uncolored vertices.
- C_k is the set of vertices of color k, on termination.

begin
  $k \longleftarrow 0,\ A \longleftarrow [n],\ U \longleftarrow [n],\ C_k \longleftarrow \emptyset.$
  while $U \neq \emptyset$ **do**
  $k \longleftarrow k+1\ A \longleftarrow U$
  while $A \neq \emptyset$
    begin
    Choose $v \in A$ and put it into C_k
    $U \longleftarrow U \setminus \{v\}$
    $A \longleftarrow A \setminus (\{v\} \cup N(v))$
    end
end

Theorem 9.9 *Suppose* $0 < p < 1$ *is a constant and* $b = \frac{1}{1-p}$. *Then w.h.p. algorithm GREEDY uses approximately* $n/\log_b n$ *colors to color the vertices of* $\mathbb{G}_{n,p}$.

Proof At the start of an iteration the edges inside U are unexamined. Suppose that

$$|U| \geq \nu = \frac{n}{(\log_b n)^2}.$$

We show that approximately $\log_b n$ vertices get color k, i.e., at the end of round k, $|C_k| \sim \log_b n$.

Each iteration chooses a *maximal* independent set from the remaining uncolored vertices. Let $k_0 = \log_b n - 5\log_b \log_b n$. Then

$$\mathbb{P}(\exists\, T : |T| \leq k_0,\, T \text{ is maximally independent in U})$$
$$\leq \sum_{t=1}^{k_0} \binom{n}{t} (1-p)^{\binom{t}{2}} \left(1-(1-p)^t\right)^{k_0-t} \leq e^{-\frac{1}{2}(\log_b n)^3}. \quad (9.24)$$

So the probability that we fail to use at least k_0 colors while $|U| \geq \nu$ is at most

$$ne^{-\frac{1}{2}(\log_b \nu)^3} = o(1).$$

So w.h.p. GREEDY uses at most

$$\frac{n}{k_0} + \nu \sim \frac{n}{\log_b n} \text{ colors.}$$

We now put a lower bound on the number of colors used by GREEDY. Let

$$k_1 = \log_b n + 2\log_b \log_b n.$$

Consider one round. Let $U_0 = U$ and suppose $u_1, u_2, \ldots \in C_k$ and $U_{i+1} = U_i \setminus (\{u_i\} \cup N(u_i))$. Then

$$\mathbb{E}(\,|U_{i+1}|\;|U_i\,) \leq |U_i\,|(1-p),$$

and so, for $i = 1, 2, \ldots$,

$$\mathbb{E}\,|U_i| \leq n(1-p)^i.$$

So

$$\mathbb{P}(k_1 \text{ vertices colored in one round}) \leq \frac{1}{(\log_b n)^2},$$

and

$$\mathbb{P}(2k_1 \text{ vertices colored in one round}) \leq \frac{1}{n}.$$

So let

$$\delta_i = \begin{cases} 1 & \text{if at most } k_1 \text{ vertices are colored in round } i, \\ 0 & \text{otherwise.} \end{cases}$$

We see that

$$\mathbb{P}(\delta_i = 1 | \delta_1, \delta_2, \ldots, \delta_{i-1}) \geq 1 - \frac{1}{(\log_b n)^2}.$$

So the number of rounds that color more than k_1 vertices is stochastically dominated by a binomial with mean $n/(\log_b n)^2$. The Chernoff bounds imply that w.h.p. the number of rounds that color more than k_1 vertices is less than $2n/(\log_b n)^2$. Strictly speaking, we need to use Lemma 2.29 to justify the use of the Chernoff bounds. Because no round colors more than $2k_1$ vertices, we see that w.h.p. GREEDY uses at least

$$\frac{n - 4k_1 n/(\log_b n)^2}{k_1} \sim \frac{n}{\log_b n} \text{ colors.}$$

□

Exercises

9.3.1 Verify equation (9.22).
9.3.2 Verify equation (9.24).

Problems for Chapter 9

9.1 Let $p = K \log n/n$ for some large constant $K > 0$. Show that w.h.p. the diameter of $\mathbb{G}_{n,p}$ is $\Theta(\log n/\log\log n)$.

9.2 Suppose that $1 + \varepsilon \leq np = o(\log n)$, where $\varepsilon > 0$ is constant. Show that given $A > 0$, there exists $B = B(A)$ such that

$$\mathbb{P}\left(\text{diam}(K) \geq B\frac{\log n}{\log np}\right) \leq n^{-A},$$

where K is the giant component of $\mathbb{G}_{n,p}$.

9.3 Complete the proof of Theorem 9.5.
Proceed as follows: let $m = d/(\log d)^2$ and partition $[n]$ into $n_0 = \frac{n}{m}$ sets $S_1, S_2, \ldots, S_{n_0}$ of size m. Let $\beta(G)$ be the maximum size of an independent set S that satisfies $|S \cap S_i| \leq 1$ for $i = 1, 2, \ldots, n_0$. Use the proof idea of Theorem 9.5 to show that w.h.p.

$$\beta(\mathbb{G}_{n,p}) \geq k_{-\varepsilon} = \frac{2n}{d}(\log d - \log\log d - \log 2 + 1 - \varepsilon).$$

9.4 Let $p = \frac{c}{n}$ where $c > 1$ is constant. Consider the greedy algorithm for constructing a large independent set I: choose a random vertex v and put v into I. Then delete v and all of its neighbors. Repeat until there are no vertices left. Show that w.h.p. this algorithm chooses an independent set of size at least $\frac{\log c}{c} n$.

9.5 Prove that if $\omega = \omega(n) \to \infty$ then there exists an interval I of length $\omega n^{1/2}/\log n$ such that w.h.p. $\chi(G_{n,1/2}) \in I$ (see Scott [107]).

9.6 A *topological clique* of size s is a graph obtained from the complete graph K_s by subdividing edges. Let $tc(G)$ denote the size of the largest topological clique contained in a graph G. Prove that w.h.p. $tc(G_{n,1/2}) = \Theta(n^{1/2})$.

9.7 Suppose that H is obtained from $G_{n,1/2}$ by planting a clique C of size $m = n^{1/2} \log n$ inside it. Describe a polynomial time algorithm that w.h.p. finds C. (Think that an adversary adds the clique without telling you where it is.)

9.8 Show that if $d > 2k \log k$ for a positive integer $k \geq 2$, then w.h.p. $G(n, d/n)$ is not k-colorable. (Hint: Consider the expected number of proper k-colorings.)

9.9 Let $p = d/n$ for some constant $d > 0$. Let A be the adjacency matrix of $G_{n,p}$. Show that w.h.p. $\lambda_1(A) \sim \Delta^{1/2}$, where Δ is the maximum degree in $G_{n,p}$. (Hint: the maximum eigenvalue of the adjacency matrix of $K_{1,m}$ is $m^{1/2}$.)

9.10 A proper *2-tone* k-coloring of a graph $G = (V, E)$ is an assignment of pairs of colors $C_v \subsetneq [k], |C_v| = 2$ such that $|C_v \cap C_w| < d(v, w)$, where $d(v, w)$ is the graph distance from v to w. If $\chi_2(G)$ denotes the minimum k for which there exists a 2-tone coloring of G, show that w.h.p. $\chi_2(\mathbb{G}_{n,p}) \sim 2\chi(\mathbb{G}_{n,p})$. (This question is taken from [10].)

9.11 Suppose that H is a graph on vertex set $[n]$ with maximum degree $\Delta = n^{o(1)}$. Let p be constant and let $G = \mathbb{G}_{n,p} + H$. Show that w.h.p. $\chi(G) \sim \chi(\mathbb{G}_{n,p})$ *for all choices of* H.

9.12 The *set chromatic number* $\chi_s(G)$ of a graph $G = (V, E)$ is defined as follows: Let C denote a set of colors. Color each $v \in V$ with a color $f(v) \in C$. Let $C_v = \{f(w) : \{v, w\} \in G\}$. The coloring is proper if $C_v \neq C_w$ whenever $\{v, w\} \in E$. χ_s is the minimum size of C in a proper coloring of G. Prove that if $0 < p < 1$ is constant, then w.h.p. $\chi_s(\mathbb{G}_{n,p}) \sim r \log_2 n$, where $r = \frac{2}{\log_2 1/s}$ and $s = \min \{q^{2\ell} + (1 - q^\ell)^2 : \ell = 1, 2, \dots\}$, where $q = 1 - p$. (This question is taken from Dudek, Mitsche and Prałat [39].)

Part III

Modeling Complex Networks

10 Inhomogeneous Graphs

Thus far, we have concentrated on the properties of the random graphs $\mathbb{G}_{n,m}$ and $\mathbb{G}_{n,p}$. These classic random graphs are not very well suited to model real-world networks. For example, the assumption in $\mathbb{G}_{n,p}$ that edges connect vertices with the same probability and independently of each other is not very realistic. To become closer to reality, we make a first modest step and consider a generalization of $\mathbb{G}_{n,p}$ where the probability of an edge $\{i, j\}$ is p_{ij} and is not the same for all pairs i, j. We call this the *generalized binomial graph*. Our main result on this model concerns the probability that it is connected.

After this we move onto a special case of this model, viz. the *expected degree model* introduced by Chung and Lu. Here p_{ij} is proportional to $w_i w_j$ for weights w_i on vertices. In this model, we prove results about the size of the largest components.

The final section introduces a tool, called the configuration model, to generate a close approximation of a random graph with a fixed degree sequence. Although initially promoted by Bollobás, this class of random graphs is sometimes called the Molloy–Reed model.

10.1 Generalized Binomial Graph

Consider the following natural generalization of the binomial random graph $\mathbb{G}_{n,p}$, first considered by Kovalenko [75].

Let $V = \{1, 2, \ldots, n\}$ be the vertex set. The random graph $\mathbb{G}_{n,\mathbf{P}}$ has vertex set V and two vertices i and j from V, $i \neq j$, are joined by an edge with probability $p_{ij} = p_{ij}(n)$, independently of all other edges. Denote by

$$\mathbf{P} = \left[p_{ij}\right]$$

the symmetric $n \times n$ matrix of edge probabilities, where $p_{ii} = 0$. Put $q_{ij} = 1 - p_{ij}$ and for $i, j \in \{1, 2, \ldots, n\}$ define

$$Q_i = \prod_{j=1}^{n} q_{ij}, \; \lambda_n = \sum_{i=1}^{n} Q_i.$$

Note that Q_i is the probability that vertex i is isolated and λ_n is the expected number of isolated vertices. Next let

$$R_{ik} = \min_{1 \le j_1 < j_2 < \cdots < j_k \le n} q_{ij_1} \cdots q_{ij_k}.$$

Suppose that the edge probabilities p_{ij} are chosen in such a way that the following conditions are simultaneously satisfied as $n \to \infty$:

$$\max_{1 \le i \le n} Q_i \to 0, \tag{10.1}$$

$$\lim_{n \to \infty} \lambda_n = \lambda = \text{constant}, \tag{10.2}$$

and

$$\lim_{n \to \infty} \sum_{k=1}^{n/2} \frac{1}{k!} \left(\sum_{i=1}^{n} \frac{Q_i}{R_{ik}} \right)^k = e^{\lambda} - 1. \tag{10.3}$$

The next two theorems are due to Kovalenko [75]. We will first give the asymptotic distribution of the number of isolated vertices in $\mathbb{G}_{n,\mathbf{P}}$, assuming that the above three conditions are satisfied. It is a generalization of the corresponding result for the classical model $\mathbb{G}_{n,p}$ (see Theorem 5.2(ii)).

Theorem 10.1 *Let X_0 denote the number of isolated vertices in the random graph $\mathbb{G}_{n,\mathbf{P}}$. If conditions (10.1), (10.2), and (10.3) hold, then*

$$\lim_{n \to \infty} \mathbb{P}(X_0 = k) = \frac{\lambda^k}{k!} e^{-\lambda}$$

for $k = 0, 1, \ldots$, i.e., the number of isolated vertices is asymptotically Poisson distributed with mean λ.

Proof Let

$$X_{ij} = \begin{cases} 1 & \text{with prob. } p_{ij}, \\ 0 & \text{with prob. } q_{ij} = 1 - p_{ij}. \end{cases}$$

Denote by X_i, for $i = 1, 2, \ldots, n$, the indicator of the event that vertex i is isolated in $\mathbb{G}_{n,\mathbf{P}}$. To show that X_0 converges in distribution to the Poisson random variable with mean λ one has to show (see Lemma 5.1) that for any natural number k,

$$\mathbb{E}\left(\sum_{1 \le i_1 < i_2 < \ldots < i_k \le n} X_{i_1} X_{i_2} \cdots X_{i_k} \right) \to \frac{\lambda^k}{k!} \tag{10.4}$$

as $n \to \infty$. But

$$\mathbb{E}\left(X_{i_1} X_{i_2} \cdots X_{i_k}\right) = \prod_{r=1}^{k} \mathbb{P}\left(X_{i_r} = 1 | X_{i_1} = \cdots = X_{i_{r-1}} = 1\right), \tag{10.5}$$

where in the case of $r = 1$ we condition on the sure event.

Since the LHS of (10.4) is the sum of $\mathbb{E}\left(X_{i_1} X_{i_2} \cdots X_{i_k}\right)$ over all $i_1 < \cdots < i_k$, we need to find matching upper and lower bounds for this expectation. Now $\mathbb{P}\left(X_{i_r}=1|X_{i_1}=\cdots=X_{i_{r-1}}=1\right)$ is the unconditional probability that i_r is not adjacent to any vertex $j\neq i_1,\ldots,i_{r-1}$ and so

$$\mathbb{P}\left(X_{i_r}=1|X_{i_1}=\cdots=X_{i_{r-1}}=1\right)=\frac{\prod_{j=1}^{n}q_{i_rj}}{\prod_{s=1}^{r-1}q_{i_ri_s}}.$$

Hence

$$Q_{i_r}\leq P\left(X_{i_r}=1|X_{i_1}=\cdots=X_{i_{r-1}}=1\right)\leq\frac{Q_{i_r}}{R_{i_r,r-1}}\leq\frac{Q_{i_r}}{R_{i_r k}}.$$

It follows from (10.5) that

$$Q_{i_1}\cdots Q_{i_k}\leq\mathbb{E}\left(X_{i_1}\cdots X_{i_k}\right)\leq\frac{Q_{i_1}}{R_{i_1k}}\cdots\frac{Q_{i_k}}{R_{i_kk}}. \tag{10.6}$$

Applying conditions (10.1) and (10.2), we get that

$$\begin{aligned}\sum_{1\leq i_1<\cdots<i_k\leq n}Q_{i_1}\cdots Q_{i_k}&=\frac{1}{k!}\sum_{1\leq i_1\neq\cdots\neq i_r\leq n}Q_{i_1}\cdots Q_{i_k}\geq\frac{1}{k!}\sum_{1\leq i_1,\ldots,i_k\leq n}Q_{i_1}\cdots Q_{i_k}\\&\quad-\frac{k}{k!}\sum_{i=1}^{n}Q_i^2\left(\sum_{1\leq i_1,\ldots,i_{k-2}\leq n}Q_{i_1}\cdots Q_{i_{k-2}}\right)\\&\geq\frac{\lambda_n^k}{k!}-(\max_i Q_i)\lambda_n^{k-1}\to\frac{\lambda^k}{k!},\end{aligned} \tag{10.7}$$

as $n\to\infty$.

Now,

$$\sum_{i=1}^{n}\frac{Q_i}{R_{ik}}\geq\lambda_n=\sum_{i=1}^{n}Q_i,$$

and if $\limsup_{n\to\infty}\sum_{i=1}^{n}\frac{Q_i}{R_{ik}}>\lambda$ then $\limsup_{n\to\infty}\sum_{k=1}^{n/2}\frac{1}{k!}\left(\sum_{i=1}^{n}\frac{Q_i}{R_{ik}}\right)^k>e^\lambda-1$, which contradicts (10.3). It follows that

$$\lim_{n\to\infty}\sum_{i=1}^{n}\frac{Q_i}{R_{ik}}=\lambda.$$

Therefore,

$$\sum_{1\leq i_1<\ldots<i_k\leq n}Q_{i_1}\cdots Q_{i_k}\leq\frac{1}{k!}\left(\sum_{i=1}^{n}\frac{Q_i}{R_{ik}}\right)^k\to\frac{\lambda^k}{k!},$$

as $n\to\infty$.

Combining this with (10.7) gives us (10.4) and completes the proof of Theorem 10.1. □

One can check (see Exercise 10.1.1) that the conditions of the theorem are satisfied when

$$p_{ij}=\frac{\log n+x_{ij}}{n},$$

where x_{ij} are uniformly bounded by a constant.

The following theorem shows that under certain circumstances, the random graph $\mathbb{G}_{n,\mathbf{P}}$ behaves in a similar way to $\mathbb{G}_{n,p}$ at the connectivity threshold.

Theorem 10.2 *If the conditions (10.1), (10.2), and (10.3) hold, then*

$$\lim_{n\to\infty} \mathbb{P}(\mathbb{G}_{n,\mathbf{P}} \textit{ is connected}) = e^{-\lambda}.$$

Proof To prove this we will show that if (10.1), (10.2), and (10.3) are satisfied then w.h.p. $\mathbb{G}_{n,\mathbf{P}}$ consists of $X_0 + 1$ connected components, i.e., $\mathbb{G}_{n,\mathbf{P}}$ consists of a single giant component plus components that are isolated vertices only. This, together with Theorem 10.1, implies the conclusion of Theorem 10.2.

Let $U \subseteq V$ be a subset of the vertex set V. We say that U is *closed* if $X_{ij} = 0$, where $X_{i,j}$ is defined in Theorem 10.1, for every i and j, where $i \in U$ and $j \in V \setminus U$. Furthermore, a closed set U is called *simple* if either U or $V \setminus U$ consist of isolated vertices only. Denote the number of nonempty closed sets in $\mathbb{G}_{n,\mathbf{P}}$ by Y_1 and the number of nonempty simple sets by Y. Clearly $Y_1 \geq Y$. We will prove first that

$$\liminf_{n\to\infty} \mathbb{E}\, Y \geq 2e^{\lambda} - 1. \tag{10.8}$$

Denote the set of isolated vertices in $\mathbb{G}_{n,\mathbf{P}}$ by J. If $V \setminus J$ is not empty then $Y = 2^{X_0+1} - 1$ (the number of nonempty subsets of J plus the number of their complements, plus V itself). If $V \setminus J = \emptyset$ then $Y = 2^n - 1$. Now, by Theorem 10.1, for every fixed $k = 0, 1, \dots,$

$$\lim_{n\to\infty} \mathbb{P}(Y = 2^{k+1} - 1) = e^{-\lambda} \frac{\lambda^k}{k!}.$$

Observe that for any $\ell \geq 0$,

$$\mathbb{E}\, Y \geq \sum_{k=0}^{\ell} (2^{k+1} - 1)\, \mathbb{P}(Y = 2^{k+1} - 1)$$

and hence

$$\liminf_{n\to\infty} \mathbb{E}\, Y \geq \sum_{k=0}^{\ell} (2^{k+1} - 1) \frac{\lambda^k e^{-\lambda}}{k!}.$$

So,

$$\liminf_{n\to\infty} \mathbb{E}\, Y \geq \lim_{\ell\to\infty} \sum_{k=0}^{\ell} (2^{k+1} - 1) \frac{\lambda^k e^{-\lambda}}{k!} = 2e^{\lambda} - 1,$$

which completes the proof of (10.8).

We will show next that

$$\limsup_{n\to\infty} \mathbb{E}\, Y_1 \leq 2e^{\lambda} - 1. \tag{10.9}$$

To prove (10.9) denote by Z_k the number of closed sets of order k in $\mathbb{G}_{n,\mathbf{P}}$ so that $Y_1 = \sum_{k=1}^{n} Z_k$. Note that

$$Z_k = \sum_{i_1<\ldots<i_k} Z_{i_1\ldots i_k},$$

where $Z_{i_1,\ldots,i_k}$ indicates whether set $I_k = \{i_1 \cdots i_k\}$ is closed. Then

$$\mathbb{E}\, Z_{i_1,\ldots,i_k} = \mathbb{P}(X_{ij} = 0, i \in I_k, j \notin I_k) = \prod_{i\in I_k, j\notin I_k} q_{ij}.$$

Consider first the case when $k \leq n/2$. Then

$$\prod_{i\in I_k, j\notin I_k} q_{ij} = \frac{\prod_{i\in I_k, 1\leq j\leq n} q_{ij}}{\prod_{i\in I_k, j\in I_k} q_{ij}} = \prod_{i\in I_k} \frac{Q_i}{\prod_{j\in I_k} q_{ij}} \leq \prod_{i\in I_k} \frac{Q_i}{R_{ik}}.$$

Hence

$$\mathbb{E}\, Z_k \leq \sum_{i_1<\ldots<i_k} \prod_{i\in I_k} \frac{Q_i}{R_{ik}} \leq \frac{1}{k!}\left(\sum_{i=1}^{n} \frac{Q_i}{R_{ik}}\right)^k.$$

Now, (10.3) implies that

$$\limsup_{n\to\infty} \sum_{k=1}^{n/2} \mathbb{E}\, Z_k \leq e^{\lambda} - 1.$$

To complete the estimation of $\mathbb{E}\, Z_k$ (and thus for $\mathbb{E}\, Y_1$), consider the case when $k > n/2$. For convenience let us switch k with $n-k$, i.e., consider $\mathbb{E}\, Z_{n-k}$, when $0 \leq k < n/2$. Notice that $\mathbb{E}\, Z_n = 1$ since V is closed. So for $1 \leq k < n/2$,

$$\mathbb{E}\, Z_{n-k} = \sum_{i_1<\ldots<i_k} \prod_{i\in I_k, j\notin I_k} q_{ij}.$$

But $q_{ij} = q_{ji}$ so, for such k, $\mathbb{E}\, Z_{n-k} = \mathbb{E}\, Z_k$. This gives

$$\limsup_{n\to\infty} \mathbb{E}\, Y_1 \leq 2(e^{\lambda} - 1) + 1,$$

where the +1 comes from $Z_n = 1$. This completes the proof of (10.9).

Now,

$$\mathbb{P}(Y_1 > Y) = \mathbb{P}(Y_1 - Y \geq 1) \leq \mathbb{E}(Y_1 - Y).$$

Estimates (10.8) and (10.9) imply that

$$\limsup_{n\to\infty} \mathbb{E}(Y_1 - Y) \leq 0,$$

which in turn leads to the conclusion that

$$\lim_{n\to\infty} \mathbb{P}(Y_1 > Y) = 0,$$

i.e., asymptotically, the probability that there is a closed set that is not simple tends to zero as $n \to \infty$. It is easy to check that $X_0 < n$ w.h.p. and therefore $Y = 2^{X_0+1} - 1$ w.h.p. and so w.h.p. $Y_1 = 2^{X_0+1} - 1$. If $\mathbb{G}_{n,\mathbf{P}}$ has more than $X_0 + 1$ connected components, then the graph after the removal of all isolated vertices would contain at least one closed set, i.e., the number of closed sets would be at least 2^{X_0+1}. But the probability of such an event tends to zero and the theorem follows. □

We finish this section by presenting a sufficient condition for $\mathbb{G}_{n,\mathbf{P}}$ to be connected w.h.p. as proven by Alon [7].

Theorem 10.3 *For every positive constant b there exists a constant $c = c(b) > 0$ so that if, for every nontrivial $S \subset V$,*

$$\sum_{i\in S, j\in V\setminus S} p_{ij} \geq c \log n,$$

then the probability that $\mathbb{G}_{n,\mathbf{P}}$ is connected is at least $1 - n^{-b}$.

Proof In fact Alon's result is much stronger. He considers a random subgraph $\mathbb{G}_{p_e}$ of a multigraph G on n vertices, obtained by deleting each edge e independently with probability $1 - p_e$. The random graph $\mathbb{G}_{n,\mathbf{P}}$ is a special case of $\mathbb{G}_{p_e}$ when G is the complete graph K_n. Therefore, following in his footsteps, we will prove that Theorem 10.3 holds for $\mathbb{G}_{p_e}$ and thus for $\mathbb{G}_{n,\mathbf{P}}$.

So, let $G = (V, E)$ be a loopless undirected multigraph on n vertices, with probability p_e, $0 \leq p_e \leq 1$, assigned to every edge $e \in E$ and suppose that for any nontrivial $S \subset V$, the expectation of the number E_S of edges in a cut $(S, V \setminus S)$ of $\mathbb{G}_{p_e}$ satisfies

$$\mathbb{E}\, E_S = \sum_{e\in(S,V\setminus S)} p_e \geq c \log n. \tag{10.10}$$

Create a new graph $G' = (V, E')$ from G by replacing each edge e by $k = c \log n$ parallel copies with the same endpoints and giving each copy e' of e a probability $p'_{e'} = p_e/k$.

Observe that for $S \subset V$,

$$\mathbb{E}\, E'_S = \sum_{e'\in(S,V\setminus S)} p'_{e'} = \mathbb{E}\, E_S.$$

Moreover, for every edge e of G, the probability that no copy e' of e survives in a random subgraph $\mathbb{G}'_{p'}$ is $(1 - p_e/k)^k \geq 1 - p_e$ and hence the probability that $\mathbb{G}_{p_e}$ is connected exceeds the probability of $\mathbb{G}'_{p'_e}$ being connected, and so in order to prove the theorem it suffices to prove that

$$\mathbb{P}(\mathbb{G}'_{p'_e} \text{ is connected}) \geq 1 - n^{-b}. \tag{10.11}$$

To prove this, let $E'_1 \cup E'_2 \cup \cdots \cup E'_k$ be a partition of the set E' of the edges of G' such that each E'_i consists of a single copy of each edge of G. For $i = 0, 1, \ldots, k$ define $\mathbb{G}'_i$ as follows. $\mathbb{G}'_0$ is the subgraph of G' that has no edges, and for all $i \geq 1$, $\mathbb{G}'_i$ is the random subgraph of G' obtained from $\mathbb{G}'_{i-1}$ by adding to it each edge $e' \in E'_i$ independently, with probability $p'_{e'}$.

Let C_i be the number of connected components of $\mathbb{G}'_i$. Then we have $C_0 = n$ and we have $\mathbb{G}'_k \equiv \mathbb{G}'_{p'_e}$. Let us call the stage i, $1 \leq i \leq k$, *successful* if either $C_{i-1} = 1$

(i.e., $\mathbb{G}'_{i-1}$ is connected) or if $C_i < 0.9C_{i-1}$. We will prove that

$$\mathbb{P}(C_{i-1} = 1 \text{ or } C_i < 0.9C_{i-1} | \mathbb{G}'_{i-1}) \geq \frac{1}{2}. \tag{10.12}$$

To see that (10.12) holds, note first that if $\mathbb{G}'_{i-1}$ is connected then there is nothing to prove. Otherwise let $\mathbb{H}_i = (U, F)$ be the graph obtained from $\mathbb{G}'_{i-1}$ by (i) contracting every connected component of $\mathbb{G}'_{i-1}$ to a single vertex and (ii) adding to it each edge $e' \in E'_i$ independently, with probability $p'_{e'}$ and throwing away loops. Note that since for every nontrivial S, $\mathbb{E}\, E'_S \geq k$, we have that for every vertex $u \in U$ (connected component of $\mathbb{G}'_{i-1}$),

$$\sum_{u \in e' \in F} p'_{e'} = \sum_{e \in E(U, U^c)} \frac{p_e}{k} \geq 1.$$

Moreover, the probability that a fixed vertex $u \in U$ is isolated in $\mathbb{H}_i$ is

$$\prod_{u \in e' \in F} (1 - p'_{e'}) \leq \exp\left\{ - \sum_{u \in e' \in F} p'_{e'} \right\} \leq e^{-1}.$$

Hence, the expected number of isolated vertices of $\mathbb{H}_i$ does not exceed $|U|e^{-1}$. Therefore, by the Markov inequality, it is at most $2|U|e^{-1}$ with probability at least 1/2. But in this case, the number of connected components of $\mathbb{H}_i$ is at most

$$2|U|e^{-1} + \frac{1}{2}(|U| - 2|U|e^{-1}) = \left(\frac{1}{2} + e^{-1}\right)|U| < 0.9|U|,$$

and so (10.12) follows. Observe that if $C_k > 1$, then the total number of successful stages is strictly less than $\log n / \log 0.9 < 10 \log n$. However, by (10.12), the probability of this event is at most the probability that a binomial random variable with parameters k and $1/2$ will attain a value at most $10 \log n$. It follows from the Chernoff–Hoeffding inequality (2.27) that if $k = c \log n = (20 + t) \log n$, then the probability that $C_k > 1$ (i.e., that $\mathbb{G}'_{p'_e}$ is disconnected) is at most $n^{-t^2/4c}$. This completes the proof of (10.11) and the theorem follows. □

Exercises

10.1.1 Check that the conditions of Theorem 10.1 are satisfied when

$$p_{ij} = \frac{\log n + x_{ij}}{n},$$

where x_{ij} are uniformly bounded by a constant.

10.1.2 Consider a special case of the generalized binomial graph called the *Kronecker random graph*, defined as a graph where each of its $n = 2^k$ vertices is a binary string of length k, for some $k \geq 1$, and between any two such vertices (strings) $\mathbf{u}, \mathbf{v}$ we put an edge independently with probability $p_{\mathbf{uv}} = \alpha^i \beta^j \gamma^{k-i-j}$, where $1 \geq \alpha \geq \beta \geq \gamma \geq 0$, and i is the number of positions (coordinates) t such that $u_t = v_t = 1$, j is the number of t where $u_t \neq v_t$, and finally $k - i - j$ is the number of t such that $u_t = v_t = 0$. Denote by $\mathbf{0}$ the vertex with all 0's. Show that when $\beta + \gamma < 1$ then vertex $\mathbf{0}$ is isolated w.h.p.

10.1.3 Show that when $\beta + \gamma = 1$ and $0 < \beta < 1$, then a random Kronecker graph defined above cannot be connected w.h.p.

10.2 Expected Degree Sequence

In this section, we will consider a special case of Kovalenko's generalized binomial model, introduced by Chung and Lu in [30], where the edge probabilities p_{ij} depend on weights assigned to vertices.

Let $V = \{1, 2, \dots, n\}$ and let w_i be the *weight* of vertex i. Now insert edges between vertices $i, j \in V$ independently with probability p_{ij} defined as

$$p_{ij} = \frac{w_i w_j}{W} \text{ where } W = \sum_{k=1}^{n} w_k.$$

We assume that $\max_i w_i^2 < W$ so that $p_{ij} \leq 1$. The resulting graph is denoted as $\mathbb{G}_{n,\mathbf{P^w}}$. Note that putting $w_i = np$ for $i \in [n]$ yields the random graph $\mathbb{G}_{n,p}$.

Notice that loops are allowed here but we will ignore them in what follows. Moreover, for vertex $i \in V$ its expected degree is

$$\sum_{j=1}^{n} \frac{w_i w_j}{W} = w_i.$$

Denote the average vertex weight by $\overline{w}$ (average expected vertex degree), i.e.,

$$\overline{w} = \frac{W}{n},$$

while, for any subset U of a vertex set V define the *volume* of U as

$$\mathrm{w}(U) = \sum_{k \in U} w_k.$$

Chung and Lu in [30] and [31] proved the following results that are summarized in the following theorem.

Theorem 10.4 *The random graph $\mathbb{G}_{n,\mathbf{P^w}}$ with a given expected degree sequence has a unique giant component w.h.p. if the expected average degree is strictly greater than 1 (i.e., $\overline{w} > 1$). Moreover, if $\overline{w} > 1$, then w.h.p. the giant component has volume*

$$\lambda_0 W + O\left(\sqrt{n}(\log n)^{3.5}\right),$$

where λ_0 is the unique nonzero root of the following equation:

$$\sum_{i=1}^{n} w_i e^{-w_i \lambda} = (1 - \lambda) \sum_{i=1}^{n} w_i.$$

Furthermore w.h.p., the second-largest component has size at most

$$(1 + o(1))\mu(\overline{w}) \log n,$$

where

$$\mu(\overline{w}) = \begin{cases} 1/(\overline{w} - 1 - \log \overline{w}) & \text{if } 1 < \overline{w} < 2, \\ 1/(1 + \log \overline{w} - \log 4) & \text{if } \overline{w} > 4/e. \end{cases}$$

Here we will prove a weaker and restricted version of the above theorem. In the current context, a giant component is one with volume $\Omega(W)$.

Theorem 10.5 *If the average expected degree $\overline{w} > 4$, then a random graph $\mathbb{G}_{n,\mathbf{P^w}}$ w.h.p. has a unique giant component and its volume is at least*

$$\left(1 - \frac{2}{\sqrt{e\overline{w}}}\right) W,$$

while the second-largest component w.h.p. has size at most

$$(1 + o(1)) \frac{\log n}{1 + \log \overline{w} - \log 4}.$$

The proof is based on a key lemma given below, proved under stronger conditions on $\overline{w}$ than in fact Theorem 10.5 requires.

Lemma 10.6 *For any positive $\varepsilon < 1$ and $\overline{w} > \frac{4}{e(1-\varepsilon)^2}$ w.h.p. every connected component in the random graph $\mathbb{G}_{n,\mathbf{P^w}}$ either has volume at least εW or has at most $\frac{\log n}{1+\log \overline{w} - \log 4 + 2\log(1-\varepsilon)}$ vertices.*

Proof We first estimate the probability of the existence of a connected component with k vertices (component of *size* k) in the random graph $\mathbb{G}_{n,\mathbf{P^w}}$. Let $S \subseteq V$ and suppose that vertices from $S = \{v_{i_1}, v_{i_2}, \ldots, v_{i_k}\}$ have respective weights $w_{i_1}, w_{i_2}, \ldots, w_{i_k}$. If the set S induces a connected subgraph of $\mathbb{G}_{n,\mathbf{P^w}}$ then it contains at least one spanning tree T. The probability of such an event equals

$$\mathbb{P}(T) = \prod_{\{v_{i_j}, v_{i_l}\} \in E(T)} w_{i_j} w_{i_l} \rho,$$

where

$$\rho := \frac{1}{W} = \frac{1}{n\overline{w}}.$$

So, the probability that S induces a connected subgraph of our random graph can be bounded from above by

$$\sum_T \mathbb{P}(T) = \sum_T \prod_{\{v_{i_j}, v_{i_l}\} \in E(T)} w_{i_j} w_{i_l} \rho,$$

where T ranges over all spanning trees on S.

By the matrix-tree theorem (see West [114]), the above sum equals the determinant of any $k-1$ by $k-1$ principal sub-matrix of $(D-A)\rho$, where A is defined as

$$A = \begin{pmatrix} 0 & w_{i_1}w_{i_2} & \cdots & w_{i_1}w_{i_k} \\ w_{i_2}w_{i_1} & 0 & \cdots & w_{i_2}w_{i_k} \\ \vdots & \vdots & \ddots & \vdots \\ w_{i_k}w_{i_1} & w_{i_k}w_{i_2} & \cdots & 0 \end{pmatrix},$$

while D is the diagonal matrix

$$D = \operatorname{diag}\left(w_{i_1}(W - w_{i_1}), \ldots, w_{i_k}(W - w_{i_k})\right).$$

(To evaluate the determinant of the first principal co-factor of $D-A$, delete row and column k of $D-A$; take out a factor $w_{i_1}w_{i_2}\cdots w_{i_{k-1}}$; add the last $k-2$ rows to row 1; Row 1 is now $(w_{i_k}, w_{i_k}, \ldots, w_{i_k})$, so we can take out a factor w_{i_k}; now subtract column 1 from the remaining columns to get a $(k-1)\times(k-1)$ upper triangular matrix with diagonal equal to $\operatorname{diag}(1, \mathrm{w}(S), \mathrm{w}(S), \ldots, \mathrm{w}(S))$).

It follows that

$$\sum_T \mathbb{P}(T) = w_{i_1}w_{i_2}\cdots w_{i_k}\mathrm{w}(S)^{k-2}\rho^{k-1}. \tag{10.13}$$

To show that this subgraph is in fact a component one has to multiply by the probability that there is no edge leaving S in $\mathbb{G}_{n,\mathbf{P}^{\mathbf{w}}}$. Obviously, this probability equals $\prod_{v_i\in S, v_j\notin S}(1 - w_iw_j\rho)$ and can be bounded from above:

$$\prod_{v_i\in S, v_j\in V\setminus S}(1 - w_iw_j\rho) \le e^{-\rho\mathrm{w}(S)(W-\mathrm{w}(S))}. \tag{10.14}$$

Let X_k be the number of components of size k in $\mathbb{G}_{n,\mathbf{P}^{\mathbf{w}}}$. Then, using the bounds from (10.13) and (10.14), we get

$$\mathbb{E}\,X_k \le \sum_S \mathrm{w}(S)^{k-2}\rho^{k-1}e^{-\rho\mathrm{w}(S)(W-\mathrm{w}(S))}\prod_{i\in S} w_i,$$

where the sum ranges over all $S \subseteq V$, $|S| = k$. Now, we focus our attention on k-vertex components whose volume is at most εW. We call such components *small* or *ε-small.* So, if Y_k is the number of small components of size k in $\mathbb{G}_{n,\mathbf{P}^{\mathbf{w}}}$, then

$$\mathbb{E}\,Y_k \le \sum_{\text{small } S} \mathrm{w}(S)^{k-2}\rho^{k-1}e^{-\mathrm{w}(S)(1-\varepsilon)}\prod_{i\in S} w_i = f(k). \tag{10.15}$$

Now, using the arithmetic-geometric mean inequality, we have

$$f(k) \le \sum_{\text{small } S}\left(\frac{\mathrm{w}(S)}{k}\right)^k \mathrm{w}(S)^{k-2}\rho^{k-1}e^{-\mathrm{w}(S)(1-\varepsilon)}.$$

The function $x^{2k-2}e^{-x(1-\varepsilon)}$ achieves its maximum at $x = (2k-2)/(1-\varepsilon)$. Therefore

$$f(k) \le \binom{n}{k}\frac{\rho^{k-1}}{k^k}\left(\frac{2k-2}{1-\varepsilon}\right)^{2k-2} e^{-(2k-2)},$$

$$\begin{aligned}
&\leq \left(\frac{ne}{k}\right)^k \frac{\rho^{k-1}}{k^k} \left(\frac{2k-2}{1-\varepsilon}\right)^{2k-2} e^{-(2k-2)}, \\
&\leq \frac{(n\rho)^k}{4\rho(k-1)^2} \left(\frac{2}{1-\varepsilon}\right)^{2k} e^{-k} \quad \text{since} \left(\frac{k-1}{k}\right)^{2k} \leq e^{-2}, \\
&= \frac{1}{4\rho(k-1)^2} \left(\frac{4}{e\overline{w}(1-\varepsilon)^2}\right)^k, \\
&= \frac{e^{-ak}}{4\rho(k-1)^2},
\end{aligned}$$

where

$$a = 1 + \log \overline{w} - \log 4 + 2\log(1-\varepsilon) > 0$$

under the assumption of Lemma 10.6.

Let $k_0 = \frac{\log n}{a}$. When k satisfies $k_0 < k < 2k_0$, we have

$$f(k) \leq \frac{1}{4n\rho(k-1)^2} = o\left(\frac{1}{\log n}\right),$$

while, when $\frac{2\log n}{a} \leq k \leq n$, we have

$$f(k) \leq \frac{1}{4n^2\rho(k-1)^2} = o\left(\frac{1}{n\log n}\right).$$

So, the probability that there exists an ε-small component of size exceeding k_0 is at most

$$\sum_{k>k_0} f(k) \leq \frac{\log n}{a} \times o\left(\frac{1}{\log n}\right) + n \times o\left(\frac{1}{n\log n}\right) = o(1).$$

This completes the proof of Lemma 10.6. □

To prove Theorem 10.5 assume that for some fixed $\delta > 0$, we have

$$\overline{w} = 4 + \delta = \frac{4}{e\,(1-\varepsilon)^2} \text{ where } \varepsilon = 1 - \frac{2}{(e\overline{w})^{1/2}} \tag{10.16}$$

and suppose that $w_1 \geq w_2 \geq \cdots \geq w_n$. Next, we show that there exists $i_0 \geq n^{1/3}$ such that

$$w_{i_0} \geq \sqrt{\frac{\left(1+\frac{\delta}{8}\right)W}{i_0}}. \tag{10.17}$$

Suppose the contrary, i.e., for all $i \geq n^{1/3}$,

$$w_i < \sqrt{\frac{\left(1+\frac{\delta}{8}\right)W}{i}}.$$

Then

$$W \leq n^{1/3}W^{1/2} + \sum_{i=n^{1/3}}^{n} \sqrt{\frac{\left(1+\frac{\delta}{8}\right)W}{i}},$$

$$\leq n^{1/3}W^{1/2} + 2\sqrt{\left(1+\frac{\delta}{8}\right)Wn}.$$

Hence,

$$W^{1/2} \leq n^{1/3} + 2\left(1+\frac{\delta}{8}\right)n^{1/2}.$$

This is a contradiction since for our choice of $\overline{w}$,

$$W = n\overline{w} \geq 4(1+\delta)n.$$

We have therefore verified the existence of i_0 satisfying (10.17).

Now consider the subgraph G of $\mathbb{G}_{n,\mathbf{P}^w}$ on the first i_0 vertices. The probability that there is an edge between vertices v_i and v_j, for any $i, j \leq i_0$, is at least

$$w_i w_j \rho \geq w_{i_0}^2 \rho \geq \frac{1+\frac{\delta}{8}}{i_0}.$$

So the asymptotic behavior of G can be approximated by a random graph $\mathbb{G}_{n,p}$ with $n = i_0$ and $p > 1/i_0$. So, w.h.p. G has a component of size $\Theta(i_0) = \Omega(n^{1/3})$. Applying Lemma 10.6 with ε as in (10.16), we see that any component with size $\gg \log n$ has volume at least εW.

Finally, consider the volume of a giant component. Suppose first that there exists a giant component of volume cW which is ε-small, i.e., $c \leq \varepsilon$. By Lemma 10.6, the size of the giant component is then at most $\frac{\log n}{2\log 2}$. Hence, there must be at least one vertex with weight w greater than or equal to the average:

$$w \geq \frac{2cW\log 2}{\log n}.$$

But it implies that $w^2 \gg W$, which contradicts the general assumption that all $p_{ij} < 1$.

We now prove uniqueness in the same way that we proved the uniqueness of the giant component in $G_{n,p}$. Choose $\eta > 0$ such that $\overline{w}(1-\eta) > 4$. Then define $w_i' = (1-\eta)w_i$ and decompose

$$\mathbb{G}_{n,\mathbf{P}^w} = G_1 \cup G_2,$$

where the edge probability in G_1 is $p'_{ij} = \frac{w'_i w'_j}{(1-\eta)W}$ and the edge probability in G_2 is p''_{ij} where $1 - \frac{w_i w_j}{W} = (1-p'_{i,j})(1-p''_{ij})$. Simple algebra gives $p''_{ij} \geq \frac{\eta w_i w_j}{W}$. It follows from the previous analysis that G_1 contains between 1 and $1/\varepsilon$ giant components. Let C_1, C_2 be two such components. The probability that there is no G_2 edge between them is at most

$$\prod_{\substack{i\in C_1\\ j\in C_2}}\left(1-\frac{\eta w_i w_j}{W}\right) \leq \exp\left\{-\frac{\eta w(C_1)w(C_2)}{W}\right\} \leq e^{-\eta W} = o(1).$$

As $1/\varepsilon < 4$, this completes the proof of Theorem 10.5. □

To add to the picture of the asymptotic behavior of the random graph $\mathbb{G}_{n,\mathbf{P}^{\mathbf{w}}}$, we will present one more result from [30]. Denote by $\overline{w^2}$ the expected second-order average degree, i.e.,

$$\overline{w^2} = \sum_j \frac{w_j^2}{W}.$$

Notice that

$$\overline{w^2} = \frac{\sum_j w_j^2}{W} \geq \frac{W}{n} = \overline{w}.$$

Chung and Lu [30] proved the following.

Theorem 10.7 *There exists a constant $C > 0$ such that if the average expected square degree $\overline{w^2} < 1$ then, with probability at least $1 - \frac{\overline{w}\left(\overline{w^2}\right)^2}{C^2\left(1-\overline{w^2}\right)}$, all components of $\mathbb{G}_{n,\mathbf{P}^{\mathbf{w}}}$ have volume at most $C\sqrt{n}$.*

Proof Let

$$x = \mathbb{P}(\exists S : w(S) \geq Cn^{1/2} \text{and } S \text{ is a component}).$$

Randomly, choose two vertices u and v from V, each with probability proportional to its weight. Then, for each vertex, the probability that it is in a set S with $w(S) \geq C\sqrt{n}$ is at least $C\sqrt{n}\rho$. Hence the probability that both vertices are in the same component is at least

$$x(C\sqrt{n}\rho)^2 = C^2xn\rho^2. \tag{10.18}$$

On the other hand, for any two fixed vertices, say u and v, the probability $P_k(u,v)$ of u and v being connected via a path of length $k+1$ can be bounded from above as follows:

$$P_k(u,v) \leq \sum_{i_1,i_2,\ldots,i_k} (w_u w_{i_1}\rho)(w_{i_1} w_{i_2}\rho)\cdots(w_{i_k} w_v\rho) \leq w_u w_v \rho(\overline{w^2})^k.$$

So the probability that u and v belong to the same component is at most

$$\sum_{k=0}^{n} P_k(u,v) \leq \sum_{k=0}^{\infty} w_u w_v \rho(\overline{w^2})^k = \frac{w_u w_v \rho}{1-\overline{w^2}}.$$

Recall that the probabilities of u and v being chosen from V are $w_u\rho$ and $w_v\rho$, respectively. Therefore the probability that a random pair of vertices are in the same component is at most

$$\sum_{u,v} w_u\rho\, w_v\rho \frac{w_u w_v \rho}{1-\overline{w^2}} = \frac{\left(\overline{w^2}\right)^2 \rho}{1-\overline{w^2}}.$$

Combining this with (10.18), we have

$$C^2xn\rho^2 \leq \frac{\left(\overline{w^2}\right)^2 \rho}{1-\overline{w^2}},$$

which implies

$$x \leq \frac{\overline{w}\left(\overline{w^2}\right)^2}{C^2\left(1-\overline{w^2}\right)},$$

and Theorem 10.7 follows. □

Exercises

10.2.1 Find the expected value and variance of the degree of a fixed vertex of a random graph $\mathbb{G}_{n,\mathbf{P^w}}$.

10.2.2 Let D_i denote the degree of vertex i in a random graph $\mathbb{G}_{n,\mathbf{P^w}}$, where $i = 1, 2, \ldots, n$. Show that w.h.p.

$$w_i - \omega(n)\sqrt{w_i} < D_i < w_i + \omega(n)\sqrt{w_i},$$

where $\omega \to \infty$ as $n \to \infty$ arbitrarily slowly.

10.2.3 Let X_0 be the number of isolated vertices in a random graph $\mathbb{G}_{n,\mathbf{P^w}}$. Show that

$$\mathbb{E}\, X_0 = \sum_{i=1}^{n} e^{-w_i} + O(\tilde{w}^2),$$

where $\tilde{w} = \overline{w^2} = \rho \sum_{i=1}^{n} w_i^2$, $\rho = 1/W$, and $W = \sum_{i=1}^{n} w_i$.

10.2.4 Let X_1 be the number of vertices of degree one in a random graph $\mathbb{G}_{n,\mathbf{P^w}}$. Show that

$$\mathbb{E}\, X_1 = \sum_{i=1}^{n} e^{-w_i} w_i + O(\tilde{w}\rho \sum_{i=1}^{n} w_i^3),$$

where $\tilde{w}$ and ρ are the same as defined in the previous exercise.

10.3 Fixed Degree Sequence

The graph $\mathbb{G}_{n,m}$ is chosen uniformly at random from the set of graphs with vertex set $[n]$ and m edges. It is of great interest to refine this model so that all the graphs chosen have a fixed degree sequence $\mathbf{d} = (d_1, d_2, \ldots, d_n)$. Of particular interest is the case where $d_1 = d_2 = \cdots = d_n = r$, i.e., the graph chosen is a uniformly random r-regular graph. It is not obvious how to do this and this is the subject of the current section.

Configuration Model

Let $\mathbf{d} = (d_1, d_2, \ldots, d_n)$ where $d_1 + d_2 + \cdots + d_n = 2m$ is even. Let

$$\mathcal{G}_{n,\mathbf{d}} = \{\text{simple graphs with vertex set } [n] \text{ s.t. degree } d(i) = d_i,\ i \in [n]\}$$

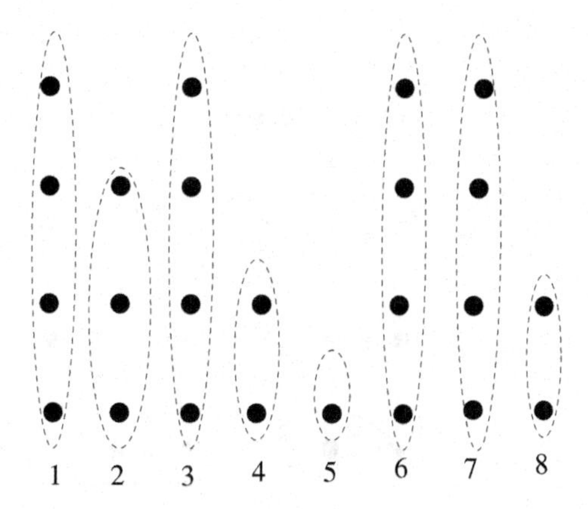

Figure 10.1 Partition of W into cells $W_1, \ldots, W_8$

and let $\mathbb{G}_{n,\mathbf{d}}$ be chosen randomly from $\mathcal{G}_{n,\mathbf{d}}$. We assume that

$$d_1, d_2, \ldots, d_n \geq 1 \text{ and } \sum_{i=1}^{n} d_i(d_i - 1) = \Omega(n).$$

We describe a generative model of $\mathbb{G}_{n,\mathbf{d}}$ due to Bollobás [18]. It is referred to as the *configuration model*. Let $W_1, W_2, \ldots, W_n$ be a partition of a set of *points* W, where $|W_i| = d_i$ for $1 \leq i \leq n$ and call the W_i *cells*. We will assume some total order $<$ on W and that $x < y$ if $x \in W_i, y \in W_j$ where $i < j$. For $x \in W$, define $\varphi(x)$ by $x \in W_{\varphi(x)}$. Let F be a partition of W into m pairs (*a configuration*). Given F we define the (multi)graph $\gamma(F)$ as

$$\gamma(F) = ([n], \{(\varphi(x), \varphi(y)) : (x, y) \in F\}).$$

Let us consider the following example of $\gamma(F)$. Let $n = 8$ and $d_1 = 4, d_2 = 3, d_3 = 4, d_4 = 2, d_5 = 1, d_6 = 4, d_7 = 4$, and $d_8 = 2$. The accompanying diagrams, Figures 10.1, 10.2, and 10.3 show a partition of W into $W_1, \ldots, W_8$, a configuration and its corresponding multigraph.

Denote by Ω the set of all configurations defined above for $d_1 + \cdots + d_n = 2m$ and notice that

$$|\Omega| = \frac{(2m)!}{m!2^m} = 1 \cdot 3 \cdot 5 \cdot \cdots \cdot (2m - 1) = (2m)!! \,. \tag{10.19}$$

To see this, take d_i "distinct" copies of i for $i = 1, 2, \ldots, n$ and take a permutation $\sigma_1, \sigma_2, \ldots, \sigma_{2m}$ of these $2m$ symbols. Read off F, pair by pair $\{\sigma_{2i-1}, \sigma_{2i}\}$ for $i = 1, 2, \ldots, m$. Each distinct F arises in $m!2^m$ ways.

We can also give an algorithmic construction of a random element F of the family Ω.

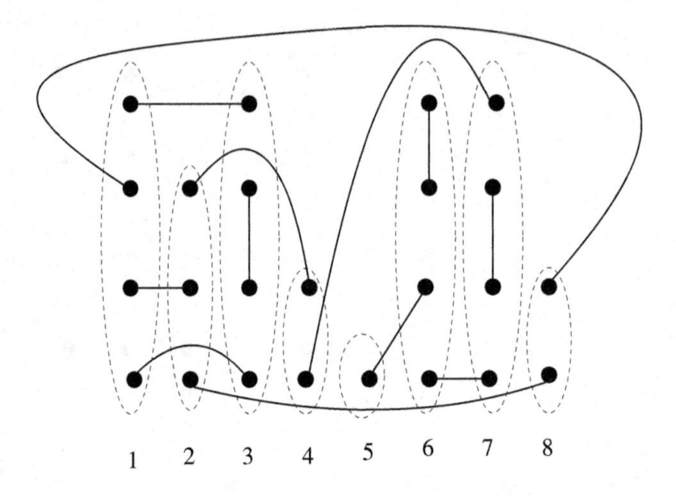

Figure 10.2 A partition F of W into $m = 12$ pairs

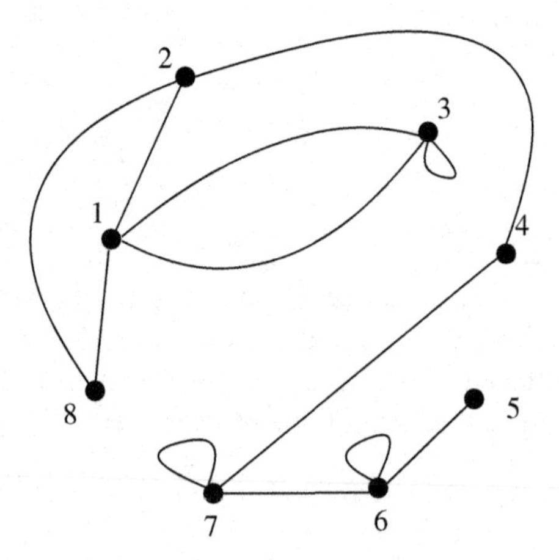

Figure 10.3 Graph $\gamma(F)$

Algorithm *F-GENERATOR*

begin
  $U \longleftarrow W,\ F \longleftarrow \emptyset$
  for $t = 1, 2, \ldots, m$ **do**
    begin
      Choose x arbitrarily from U;
      Choose y randomly from $U \setminus \{x\}$;
      $F \longleftarrow F \cup \{(x, y)\}$;
      $U \longleftarrow U \setminus \{(x, y)\}$
    end
end

Note that F arises with probability $1/[(2m-1)(2m-3)\cdots 1] = |\Omega|^{-1}$.

Observe the following relationship between a simple graph $G \in \mathcal{G}_{n,\mathbf{d}}$ and the number of configurations F for which $\gamma(F) = G$.

Lemma 10.8 *If $G \in \mathcal{G}_{n,\mathbf{d}}$, then*

$$|\gamma^{-1}(G)| = \prod_{i=1}^{n} d_i! \,.$$

Proof Arrange the edges of G in lexicographic order. Now go through the sequence of $2m$ symbols, replacing each i by a new member of W_i. We obtain all F for which $\gamma(F) = G$. □

The above lemma implies that we can use random configurations to "approximate" random graphs with a given degree sequence.

Corollary 10.9 *If F is chosen uniformly at random from the set of all configurations Ω and $G_1, G_2 \in \mathcal{G}_{n,\mathbf{d}}$, then*

$$\mathbb{P}(\gamma(F) = G_1) = \mathbb{P}(\gamma(F) = G_2).$$

So instead of sampling from the family $\mathcal{G}_{n,\mathbf{d}}$ and counting graphs with a given property, we can choose a random F and accept $\gamma(F)$ if and only if there are no loops or multiple edges, i.e., if and only if $\gamma(F)$ is a simple graph.

This is only a useful exercise if $\gamma(F)$ is simple with sufficiently high probability. We will assume for the remainder of this section that

$$\Delta = \max\{d_1, d_2, \ldots, d_n\} = O(1).$$

Lemma 10.10 *If F is chosen uniformly (at random) from Ω,*

$$\mathbb{P}(\gamma(F) \text{ is simple}) = (1 + o(1))e^{-\lambda(\lambda+1)}, \tag{10.20}$$

where

$$\lambda = \frac{\sum d_i(d_i - 1)}{2\sum d_i}. \tag{10.21}$$

Proof Notice that in order to estimate the probability that $\gamma(F)$ is simple, we can restrict our attention to the occurrence of loops and double edges only. So, let L^* denote the number of loops and let D^* be the number of double edges in $\gamma(F)$.

We first argue that for $k = O(1)$,

$$\mathbb{E}\left(\binom{L^*}{k}\right) = \sum_{\substack{S\subseteq[n]\\|S|=k}} \prod_{i\in S} \frac{d_i(d_i-1)}{4m-O(1)} \tag{10.22}$$
$$= \frac{1}{k!}\left(\sum_{i=1}^{n} \frac{d_i(d_i-1)}{4m}\right)^k + O\left(\frac{\Delta^4}{m}\right)$$
$$\sim \frac{\lambda^k}{k!}.$$

Explanation for (10.22)**:** We assume that F-GENERATOR begins with pairing up points in S. Therefore the random choice here is always from a set of size $2m - O(1)$.

It follows from Lemma 5.1 that the number of loops L^* in $\gamma(F)$ is asymptotically Poisson and hence that

$$\mathbb{P}(L^* = 0) \sim e^{-\lambda}. \tag{10.23}$$

Now, consider double edges in $\gamma(F)$ and split the set of all D^* double edges into two subsets: *non-adjacent* and *adjacent* ones, of cardinalities D_1^* and D_2^*, respectively. Obviously $D^* = D_1^* + D_2^*$.

We first show that D_1^* is asymptotically Poisson and asymptotically independent of L^*. So, let $k = O(1)$. If $\mathcal{D}_k$ denotes the set of collections of $2k$ configuration points making up k double edges, then

$$\mathbb{E}\left(\binom{D_1^*}{k}\middle| L^* = 0\right) = \sum_{\mathcal{D}_k} \mathbb{P}(\mathcal{D}_k \subseteq F \mid L^* = 0)$$
$$= \sum_{\mathcal{D}_k} \frac{\mathbb{P}(L^* = 0 \mid \mathcal{D}_k \subseteq F)\,\mathbb{P}(\mathcal{D}_k \subseteq F)}{\mathbb{P}(L^* = 0)}.$$

Now because $k = O(1)$, we see that the calculations that give us (10.23) will give us $\mathbb{P}(L^* = 0 \mid \mathcal{D}_k \subseteq F) \sim \mathbb{P}(L^* = 0)$. So,

$$\mathbb{E}\left(\binom{D_1^*}{k}\middle| L^* = 0\right) \sim \sum_{\mathcal{D}_k} \mathbb{P}(\mathcal{D}_k \subseteq F)$$
$$= \frac{1}{2} \sum_{\substack{S,T\subseteq[n]\\|S|=|T|=k\\S\cap T=\emptyset}} \sum_{\varphi:S\to T} \prod_{i\in S} \frac{2\binom{d_i}{2}\binom{d_{\varphi(i)}}{2}}{(2m-O(1))^2}$$
$$= \frac{1}{k!}\left(\sum_{i=1}^{n} \frac{d_i(d_i-1)}{4m}\right)^{2k} + O\left(\frac{\Delta^8}{m}\right)$$
$$\sim \frac{\lambda^{2k}}{k!}.$$

It follows from Lemma 5.1 that

$$\mathbb{P}(D_1^* = 0 \mid L^* = 0) \sim e^{-\lambda^2}. \tag{10.24}$$

To complete the proof of the Lemma, we need to show that w.h.p. there are no adjacent double edges. If D_2^* is the number of adjacent double edges, then

$$\mathbb{E}(D_2^*) \leq \sum_{i,j,k} \frac{d_i^4 d_j^2 d_k^2}{(2m - O(1))^4} = O(n^{-1}). \tag{10.25}$$

Explanation: We sum over the possibilities of a double edge $\{i, j\}$ and a double edge $\{i, k\}$. Having chosen vertices i, j, k, we choose two ordered pairs u_1, u_2 and v_1, v_2 from W_i in $4!\binom{d_i}{4}/2 < d_i^4$ ways; an ordered pair x_1, x_2 from W_j in less than d_j^2 ways; and an ordered pair y_1, y_2 from W_k in less than d_k^2 ways. We then estimate the probability that our configuration contains the pairs $\{u_1, x_1\}, \{u_2, x_2\}, \{v_1, y_1\}, \{v_2, y_2\}$. Using F-GENERATOR with initial choices u_1, u_2, v_1, v_2 for x, we see that the probability that u_1 is paired with x_1 is $1/(2m - 1)$. Given this, the probability that u_2 is paired with x_2 is $1/(2m - 3)$, and so on.

In conclusion, (10.25) implies that indeed w.h.p. there are no adjacent double edges and the lemma follows from (10.23) and (10.24). □

Theorem 10.11 *Suppose that* $\Delta = O(1)$.

$$|\mathcal{G}_{n,\mathbf{d}}| \sim e^{-\lambda(\lambda+1)} \frac{(2m)!!}{\prod_{i=1}^n d_i!},$$

where λ *is defined in* (10.21).

Proof This follows from Lemma 10.8 and Lemma 10.10. □

Hence, (10.19) and (10.20) will tell us not only how large $\mathcal{G}_{n,\mathbf{d}}$ is (Theorem 10.11) but also lead to the following conclusion.

Theorem 10.12 *Suppose that* $\Delta = O(1)$. *For any (multi)graph property* $\mathcal{P}$,

$$\mathbb{P}(\mathbb{G}_{n,\mathbf{d}} \in \mathcal{P}) \leq (1 + o(1))e^{\lambda(\lambda+1)}\, \mathbb{P}(\gamma(F) \in \mathcal{P}).$$

Existence of a Giant Component

Molloy and Reed [90] provide an elegant and very useful criterion for when $\mathbb{G}_{n,\mathbf{d}}$ has a giant component. Suppose that there are $\lambda_i n + o(n^{3/4})$ vertices of degree $i = 1, 2, \ldots, L$. We will assume that $L = O(1)$ and that the $\lambda_i, i \in [L]$ are constants independent of n. The paper [90] allows for $L = O(n^{1/4-\varepsilon})$. We will assume that $\lambda_1 + \lambda_2 + \cdots + \lambda_L = 1$.

Theorem 10.13 *Let* $\Lambda = \sum_{i=1}^L i(i-2)\lambda_i$. *Let* $\varepsilon > 0$ *be arbitrary.*

(a) If $\Lambda < -\varepsilon$, *then w.h.p. the size of the largest component in* $\mathbb{G}_{n,\mathbf{d}}$ *is* $O(\log n)$.
(b) If $\Lambda > \varepsilon$, *then w.h.p. there is a unique giant component of linear size* $\sim \Theta n$ *where* Θ *is defined as follows: let* $K = \sum_{i=1}^L i\lambda_i$ *and*

$$f(\alpha) = K - 2\alpha - \sum_{i=1}^{L} i\lambda_i \left(1 - \frac{2\alpha}{K}\right)^{i/2}. \tag{10.26}$$

Let ψ be the smallest positive solution to $f(\alpha) = 0$. Then

$$\Theta = 1 - \sum_{i=1}^{L} \lambda_i \left(1 - \frac{2\psi}{K}\right)^{i/2}.$$

If $\lambda_1 = 0$ then $\Theta = 1$, otherwise $0 < \Theta < 1$.

(c) *In case (b), the degree sequence of the graph obtained by deleting the giant component satisfies the conditions of (a).*

To prove the above theorem, we need Azuma–Hoeffding bounds for martingales. Recall that a *martingale* is a sequence of random variables $X_0, X_1, \ldots, X_n$ on a probability space $(\Omega, \mathcal{F}, \mathbb{P})$ with $\mathbb{E}(X_{k+1}|\mathcal{F}_k) = X_k$, where

$$\mathcal{F}_0 = \{\emptyset, \Omega\} \subseteq \mathcal{F}_1 \subseteq \mathcal{F}_2 \subseteq \cdots \subseteq \mathcal{F}_n = \mathcal{F}$$

is an increasing sequence of σ-fields. When $\mathbb{E}(X_{k+1}|\mathcal{F}_k) \geq X_k$ then this sequence is called a *super-martingale*, while if $\mathbb{E}(X_{k+1}|\mathcal{F}_k) \leq X_k$ then it is called a *sub-martingale*.

Lemma 10.14 (Azuma–Hoeffding bound) *Let $\{X_k\}_0^n$ be a sequence of random variables such that $|X_k - X_{k-1}| \leq c_k$, $k = 1, \ldots, n$, and X_0 is constant.*

(i) *If $\{X_k\}_0^n$ is a martingale, then for all $t > 0$, we have*

$$\mathbb{P}(|X_n - X_0| \geq t) \leq 2\exp\left\{-\frac{t^2}{2\sum_{i=1}^n c_i^2}\right\},$$

$$\mathbb{P}(X_n \leq \mathbb{E}\, X_n - t) \leq \exp\left\{-\frac{t^2}{2\sum_{i=1}^n c_i^2}\right\}.$$

(ii) *If $\{X_k\}_0^n$ is a super-martingale, then for all $t > 0$, we have*

$$\mathbb{P}(X_n \geq X_0 + t) \leq \exp\left\{-\frac{t^2}{2\sum_{i=1}^n c_i^2}\right\}.$$

(iii) *If $\{X_k\}_0^n$ is a sub-martingale, then for all $t > 0$, we have*

$$\mathbb{P}(X_n \leq X_0 - t) \leq \exp\left\{-\frac{t^2}{2\sum_{i=1}^n c_i^2}\right\}.$$

For a proof of Lemma 10.14, see, for example, section 21.7 of the book [52].

Proof (of Theorem 10.13)
We consider the execution of F-GENERATOR. We keep a sequence of partitions $U_t, A_t, E_t, t = 1, 2, \ldots, m$, of W. Initially $U_0 = W$ and $A_0 = E_0 = \emptyset$. The $(t+1)$th iteration of F-GENERATOR is now executed as follows: it is designed so that we construct $\gamma(F)$ component by component. A_t is the set of points associated with the *partially exposed* vertices of the current component. These are vertices in the

current component, not all of whose points have been paired. U_t is the set of unpaired points associated with the *entirely unexposed* vertices that have not been added to any component so far. E_t is the set of paired points. Whenever possible, we choose to make a pairing that involves the current component.

(i) If $A_t = \emptyset$ then choose x from U_t. Go to (iii).
We begin the exploration of a new component of $\gamma(F)$.
(ii) If $A_t \neq \emptyset$ choose x from A_t. Go to (iii).
Choose a point associated with a partially exposed vertex of the current component.
(iii) Choose y randomly from $(A_t \cup U_t) \setminus \{x\}$.
(iv) $F \leftarrow F \cup \{(x, y)\}$; $E_{t+1} \leftarrow E_t \cup \{x, y\}$; $A_{t+1} \leftarrow A_t \setminus \{x\}$.
(v) If $y \in A_t$ then $A_{t+1} \leftarrow A_{t+1} \setminus \{y\}$; $U_{t+1} \leftarrow U_t$.
y is associated with a vertex in the current component.
(vi) If $y \in U_t$ then $A_{t+1} \leftarrow A_t \cup (W_{\varphi(y)} \setminus y)$; $U_{t+1} \leftarrow U_t \setminus W_{\varphi(y)}$.
y is associated with a vertex $v = \varphi(y)$ not in the current component. Add all the points in $W_v \setminus \{y\}$ to the active set.
(vii) Go to (i).

(a) We fix a vertex v and estimate the size of the component containing v. We keep track of the size of A_t for $t = O(\log n)$ steps. Observe that

$$\mathbb{E}(|A_{t+1}| - |A_t| \mid |A_t| > 0) \lesssim \frac{\sum_{i=1}^{L} i\lambda_i n(i-2)}{M_1 - 2t - 1} = \frac{\Lambda n}{M_1 - 2t - 1} \leq -\frac{\varepsilon}{L}. \tag{10.27}$$

Here $M_1 = \sum_{i=1}^{L} i\lambda_i n$. The explanation for (10.27) is that $|A|$ increases only in Step (vi) and there it increases by $i - 2$ with probability $\lesssim \frac{i\lambda_i n}{M_1 - 2t}$. The two points x, y are missing from A_{t+1} and this explains the -2 term.

Let $\varepsilon_1 = \varepsilon/L$ and let

$$Y_t = \begin{cases} |A_t| + \varepsilon_1 t & |A_1|, |A_2|, \ldots, |A_t| > 0, \\ 0 & \text{otherwise.} \end{cases}$$

It follows from (10.27) that if $t = O(\log n)$ and $Y_1, Y_2, \ldots, Y_t > 0$, then

$$\mathbb{E}(Y_{t+1} \mid Y_1, Y_2, \ldots, Y_t) = \mathbb{E}(|A_{t+1}| + \varepsilon_1(t+1) \mid Y_1, Y_2, \ldots, Y_t) \leq |A_t| + \varepsilon_1 t = Y_t.$$

Otherwise, $\mathbb{E}(Y_{t+1} \mid \cdot) = 0 = Y_t$. It follows that the sequence (Y_t) is a super-martingale. Next let $Z_1 = 0$ and $Z_t = Y_t - Y_{t-1}$ for $t \geq 1$. Then, we have $-2 \leq Z_i \leq L$ and $\mathbb{E}(Z_i) \leq -\varepsilon_1$ for $i = 1, 2, \ldots, t$. Now,

$$\mathbb{P}(A_\tau \neq \emptyset, 1 \leq \tau \leq t) \leq \mathbb{P}(Y_t = Z_1 + Z_2 + \cdots + Z_t > 0).$$

It follows from Lemma 10.14 that if $Z = Z_1 + Z_2 + \cdots + Z_t$, then

$$\mathbb{P}(Z > 0) \leq P(Z - \mathbb{E}(Z) \geq t\varepsilon_1) \leq \exp\left\{-\frac{\varepsilon_1^2 t^2}{8t}\right\}.$$

It follows that with probability $1 - O(n^{-2})$, A_t will become empty after at most $16\varepsilon_1^{-2} \log n$ rounds. Thus for any fixed vertex v, with probability $1 - O(n^{-2})$ the component containing v has size at most $4\varepsilon_1^{-2} \log n$. (We can expose the component containing v through our choice of x in Step (i).) Thus, the probability that there is a component of size greater than $16\varepsilon_1^{-2} \log n$ is $O(n^{-1})$. This completes the proof of (a).
(b) If $t \leq \delta n$ for a small positive constant $\delta \ll \varepsilon/L^3$, then

$$\mathbb{E}(|A_{t+1}| - |A_t|) \geq \frac{-2|A_t| + (1+o(1)) \sum_{i=1}^{L} i(\lambda_i n - 2t)(i-2)}{M_1 - 2\delta n} \geq \frac{-2L\delta n + (1+o(1))(\Lambda n - 2\delta L^3 n)}{M_1 - 2\delta n} \geq \frac{\varepsilon}{2L}. \tag{10.28}$$

Let $\varepsilon_2 = \varepsilon/2L$ and let

$$Y_t = \begin{cases} |A_t| - \varepsilon_2 t & |A_1|, |A_2|, \ldots, |A_t| > 0, \\ 0 & \text{otherwise.} \end{cases}$$

It follows from (10.27) that if $t \leq \delta n$ and $Y_1, Y_2, \ldots, Y_t > 0$, then

$$\mathbb{E}(Y_{t+1} \mid Y_1, Y_2, \ldots, Y_t) = \mathbb{E}(|A_{t+1}| - \varepsilon_2(t+1) \mid Y_1, Y_2, \ldots, Y_t) \geq |A_t| - \varepsilon_2 t = Y_t.$$

Otherwise, $\mathbb{E}(Y_{t+1} \mid \cdot) = 0 = Y_t$. It follows that the sequence (Y_t) is a sub-martingale. Next let $Z_1 = 0$ and $Z_t = Y_t - Y_{t-1}$ for $t \geq 1$. Then, we have (i) $-2 \leq Z_i \leq L$ and (ii) $\mathbb{E}(Z_i) \geq \varepsilon_2$ for $i = 1, 2, \ldots, t$. Now,

$$\mathbb{P}(A_t \neq \emptyset) \geq \mathbb{P}(Y_t = Z_1 + Z_2 + \cdots + Z_t > 0).$$

It follows from Lemma 10.14 that if $Z = Z_1 + Z_2 + \cdots + Z_t$, then

$$\mathbb{P}(Z \leq 0) \leq P(Z - \mathbb{E}(Z) \geq t\varepsilon_2) \leq \exp\left\{-\frac{\varepsilon_2^2 t^2}{2t}\right\}.$$

It follows that if $L_0 = 100\varepsilon_2^{-2}$, then

$$\begin{aligned} &\mathbb{P}\left(\exists L_0 \log n \leq t \leq \delta n : Z \leq \frac{\varepsilon_2 t}{2}\right) \\ &\quad \leq \mathbb{P}\left(\exists L_0 \log n \leq t \leq \delta n : Z - \mathbb{E}(Z) \geq \frac{\varepsilon_2 t}{2}\right) \\ &\quad \leq n \exp\left\{-\frac{\varepsilon_2^2 L_0 \log n}{8}\right\} = O(n^{-2}). \end{aligned}$$

It follows that if $t_0 = \delta n$ then w.h.p. $|A_{t_0}| = \Omega(n)$, and there is a giant component and that the edges exposed between time $L_0 \log n$ and time t_0 are part of exactly one giant.

We now deal with the special case where $\lambda_1 = 0$. There are two sub-cases. If in addition we have $\lambda_2 = 1$, then w.h.p. $\mathbb{G}_{\mathbf{d}}$ is the union of $O(\log n)$ vertex disjoint cycles, see Problem 10.7. If $\lambda_1 = 0$ and $\lambda_2 < 1$, then the only solutions to $f(\alpha) = 0$ are $\alpha = 0, K/2$. For then $0 < \alpha < K/2$ implies

$$\sum_{i=2}^{L} i\lambda_i \left(1 - \frac{2\alpha}{K}\right)^{i/2} < \sum_{i=2}^{L} i\lambda_i \left(1 - \frac{2\alpha}{K}\right) = K - 2\alpha.$$

This gives $\Theta = 1$. Exercise 10.5.2 asks for a proof that w.h.p. in this case, $\mathbb{G}_{n,\mathbf{d}}$ consists of a giant component plus a collection of small components that are cycles of size $O(\log n)$.

Assume now that $\lambda_1 > 0$. We show that w.h.p. there are $\Omega(n)$ isolated edges. This together with the rest of the proof implies that $\Psi < K/2$ and hence that $\Theta < 1$. Indeed, if Z denotes the number of components that are isolated edges, then

$$\mathbb{E}(Z) = \binom{\lambda_1 n}{2} \frac{1}{2M_1 - 1} \text{ and } \mathbb{E}(Z(Z-1)) = \binom{\lambda_1 n}{4} \frac{6}{(2M_1 - 1)(2M_1 - 3)}$$

and so the Chebyshev inequality (2.18) implies that $Z = \Omega(n)$ w.h.p.

Now for i such that $\lambda_i > 0$, we let $X_{i,t}$ denote the number of entirely unexposed vertices of degree i. We focus on the number of unexposed vertices of a given degree. Then,

$$\mathbb{E}(X_{i,t+1} - X_{i,t}) = -\frac{iX_{i,t}}{M_1 - 2t - 1}. \tag{10.29}$$

This suggests that the random variables $X_{i,t}/n$ closely follow the trajectory of the differential equation, where $\tau = t/n$ and $x(\tau) = X_{i,t}/n$. This is the basis of the differential equations method and a rigorous treatment of this case can be found in Section 11.3 of [52]. We have

$$\frac{dx}{d\tau} = -\frac{ix}{K - 2\tau}. \tag{10.30}$$

$x(0) = \lambda_i$. Note that $K = M_1/n$.

The solution to (10.30) is

$$x = \lambda_i \left(1 - \frac{2\tau}{K}\right)^{i/2}. \tag{10.31}$$

Theorem 23.1 of [52] then implies that with probability $1 - O(n^{1/4} e^{-\Omega(n^{1/4})})$,

$$\left| X_{i,t} - ni\lambda_i \left(1 - \frac{2t}{K}\right)^{i/2} \right| = O(n^{3/4}), \tag{10.32}$$

up to a point where $X_{i,t} = O(n^{3/4})$. (The $O(n^{3/4})$ term for the number of vertices of degree i is absorbed into the RHS of (10.32).)

Now because

$$|A_t| = M_1 - 2t - \sum_{i=1}^{L} iX_{i,t} = Kn - 2t - \sum_{i=1}^{L} iX_{i,t},$$

we see that w.h.p.

$$\begin{aligned} |A_t| &= n\left(K - \frac{2t}{n} - \sum_{i=1}^{L} i\lambda_i \left(1 - \frac{2t}{Kn}\right)^{i/2}\right) + O(n^{3/4}) \\ &= nf\left(\frac{t}{n}\right) + O(n^{3/4}), \end{aligned} \tag{10.33}$$

so that w.h.p. the first time after time $t_0 = \delta n$ that $|A_t| = O(n^{3/4})$ is at time $t_1 = \Psi n + O(n^{3/4})$. This shows that w.h.p. there is a component of size at least

$\Theta n + O(n^{3/4})$. Indeed, we simply subtract the number of entirely unexposed vertices from n to obtain this.

To finish, we must show that this component is unique and no larger than $\Theta n + O(n^{3/4})$. We can do this by proving (c), i.e. showing that the degree sequence of the graph $\mathbb{G}_U$ induced by the unexposed vertices satisfies the condition of Case (a). For then by Case (a), the giant component can only add $O(n^{3/4} \times \log n) = o(n)$ vertices from t_1 onwards.

We first observe that the above analysis shows that w.h.p. the degree sequence of $\mathbb{G}_U$ is asymptotically equal to $n\lambda_i'$, $i = 1, 2, \ldots, L$, where

$$\lambda_i' = \lambda_i \left(1 - \frac{2\Psi}{K}\right)^{i/2}.$$

(The important thing here is that the number of vertices of degree i is asymptotically proportional to λ_i'.) Next choose $\varepsilon_1 > 0$ sufficiently small and let $t_{\varepsilon_1} = \max\{t : |A_t| \geq \varepsilon_1 n\}$. There must exist $\varepsilon_2 < \varepsilon_1$ such that $t_{\varepsilon_1} \leq (\Psi - \varepsilon_2)n$ and $f'(\Psi - \varepsilon_2) \leq -\varepsilon_1$, else f cannot reach zero. Recall that $\Psi < K/2$ here and then,

$$\begin{aligned}
-\varepsilon_1 \geq f'(\Psi - \varepsilon_2) &= -2 + \frac{1}{K - 2(\Psi - \varepsilon_2)} \sum_{i \geq 1} i^2 \lambda_i \left(1 - \frac{2\Psi - 2\varepsilon_2}{K}\right)^{i/2} \\
&= -2 + \frac{1 + O(\varepsilon_2)}{K - 2\Psi} \sum_{i \geq 1} i^2 \lambda_i \left(1 - \frac{2\Psi}{K}\right)^{i/2} \\
&= \frac{1 + O(\varepsilon_2)}{K - 2\Psi} \left(-2 \sum_{i \geq 1} i \lambda_i \left(1 - \frac{2\Psi}{K}\right)^{i/2} + \sum_{i \geq 1} i^2 \lambda_i \left(1 - \frac{2\Psi}{K}\right)^{i/2}\right) \\
&= \frac{1 + O(\varepsilon_2)}{K - 2\Psi} \sum_{i \geq 1} i(i-2) \lambda_i \left(1 - \frac{2\Psi}{K}\right)^{i/2} \\
&= \frac{1 + O(\varepsilon_2)}{K - 2\Psi} \sum_{i \geq 1} i(i-2) \lambda_i'. \qquad (10.34)
\end{aligned}$$

This completes the proofs of (b) and (c). □

Connectivity of Regular Graphs

Bollobás [18] used the configuration model to prove the following: Let $\mathbb{G}_{n,r}$ denote a random r-regular graph with vertex set $[n]$ and $r \geq 3$ constant.

Theorem 10.15 *$\mathbb{G}_{n,r}$ is r-connected, w.h.p.*

Since an r-regular, r-connected graph, with n even, has a perfect matching, the above theorem immediately implies the following corollary.

Corollary 10.16 *Let $\mathbb{G}_{n,r}$ be a random r-regular graph, $r \geq 3$ constant, with vertex set $[n]$ even. Then w.h.p. $\mathbb{G}_{n,r}$ has a perfect matching.*

Proof (of Theorem 10.15)
Partition the vertex set $V = [n]$ of $\mathbb{G}_{n,r}$ into three parts, K, L, and $V \setminus (K \cup L)$ such that $L = N(K)$, i.e., such that L separates K from $V \setminus (K \cup L)$ and $|L| = l \leq r - 1$. We will show that w.h.p. there are no such K, L for k ranging from 2 to $n/2$. We will use the configuration model and the relationship stated in Theorem 10.12. We will divide the whole range of k into three parts.

(i) $2 \leq k \leq 3$.

Put $S := K \cup L,\ s = |S| = k + l \leq r + 2$. The set S contains at least $2r - 1$ edges ($k = 2$) or at least $3r - 3$ edges ($k = 3$). In both cases this is at least $s + 1$ edges.

$$
\begin{aligned}
\mathbb{P}(\exists S, s = |S| \leq r + 2 :\ & S \text{ contains } s + 1 \text{ edges}) \\
&\leq \sum_{s=4}^{r+2} \binom{n}{s} \binom{rs}{s+1} \left(\frac{rs}{rn}\right)^{s+1} \qquad (10.35) \\
&\leq \sum_{s=4}^{r+2} n^s 2^{rs} s^{s+1} n^{-s-1} \\
&= o(1).
\end{aligned}
$$

Explanation for (10.35)**:** Having chosen a set of s vertices, spanning rs points R, we choose $s + 1$ of these points T. $\frac{rs}{rn}$ bounds the probability that one of these points in T is paired with something in a cell associated with S. This bound holds conditional on other points of R being so paired.

(ii) $4 \leq k \leq ne^{-10}$.

The number of edges incident with the set K, $|K| = k$, is at least $(rk + l)/2$. Indeed, let $a = e(K)$ and $b = e(K, L)$ edges. Then $2a + b = rk$ and $b \geq l$. This gives $a + b \geq (rk + l)/2$. So,

$$
\begin{aligned}
\mathbb{P}(\exists K, L) &\leq \sum_{k=4}^{ne^{-10}} \sum_{l=0}^{r-1} \binom{n}{k} \binom{n}{l} \binom{rk}{\frac{rk+l}{2}} \left(\frac{r(k+l)}{rn}\right)^{(rk+l)/2} \\
&\leq \sum_{k=4}^{ne^{-10}} \sum_{l=0}^{r-1} n^{-\left(\frac{r}{2}-1\right)k+\frac{l}{2}} \frac{e^{k+l}}{k^k l^l} 2^{rk} (k+l)^{(rk+l)/2}.
\end{aligned}
$$

Now

$$
\left(\frac{k+l}{l}\right)^{l/2} \leq e^{k/2} \text{ and } \left(\frac{k+l}{k}\right)^{k/2} \leq e^{l/2},
$$

and so

$$
(k+l)^{(rk+l)/2} \leq l^{l/2} k^{rk/2} e^{(lr+k)/2}.
$$

Therefore, with C_r a constant,

$$\mathbb{P}(\exists K, L) \leq C_r \sum_{k=4}^{ne^{-10}} \sum_{l=0}^{r-1} n^{-(\frac{r}{2}-1)k+\frac{l}{2}} e^{3k/2} 2^{rk} k^{(r-2)k/2}$$
$$= C_r \sum_{k=4}^{ne^{-10}} \sum_{l=0}^{r-1} \left(n^{-(\frac{r}{2}-1)+\frac{l}{2k}} e^{3/2} 2^r k^{\frac{r}{2}-1}\right)^k$$
$$= o(1). \tag{10.36}$$

(iii) $ne^{-10} < k \leq n/2$.

Assume that there are a edges between sets L and $V \setminus (K \cup L)$. Define

$$\varphi(2m) = \frac{(2m)!}{m!\, 2^m} \sim 2^{1/2} \left(\frac{2m}{e}\right)^m.$$

Then, remembering that $r, l, a = O(1)$, we can estimate that

$$\mathbb{P}(\exists K, L)$$
$$\leq \sum_{k,l,a} \binom{n}{k}\binom{n}{l}\binom{rl}{a} \frac{\varphi(rk+rl-a)\varphi(r(n-k-l)+a)}{\varphi(rn)} \tag{10.37}$$
$$\leq C_r \sum_{k,l,a} \left(\frac{ne}{k}\right)^k \left(\frac{ne}{l}\right)^l$$
$$\times \frac{(rk+rl-a)^{rk+rl-a}(r(n-k-l)+a)^{r(n-k-l)+a}}{(rn)^{rn}}$$
$$\leq C_r' \sum_{k,l,a} \left(\frac{ne}{k}\right)^k \left(\frac{ne}{l}\right)^l \frac{(rk)^{rk+rl-a}(r(n-k))^{r(n-k-l)+a}}{(rn)^{rn}}$$
$$\leq C_r'' \sum_{k,l} \left(\frac{ne}{k}\right)^k \left(\frac{ne}{l}\right)^l \left(\frac{k}{n}\right)^{rk} \left(1-\frac{k}{n}\right)^{r(n-k)}$$
$$\leq C_r''' \sum_{k} \left(\left(\frac{k}{n}\right)^{r-1} e^{1-r/2} n^{r/k}\right)^k$$
$$= o(1). \tag{10.38}$$

Explanation of (10.37)**:** Having chosen K, L we choose a points in $W_{K\cup L} = \bigcup_{i\in K\cup L} W_i$ that will be paired outside $W_{K\cup L}$. This leaves $rk + rl - a$ points in $W_{K\cup L}$ to be paired up in $\varphi(rk + rl - a)$ ways and then the remaining points can be paired up in $\varphi(r(n-k-l)+a)$ ways. We then multiply by the probability $1/\varphi(rn)$ of the final pairing. □

Exercises

10.3.1 Verify equation (10.36).

10.3.2 Verify equation (10.38).

Problems for Chapter 10

10.1 Prove that the Kronecker random graph, defined in Exercise 10.1.2, is connected w.h.p. (for $k \to \infty$) if either (i) $\beta + \gamma > 1$ or (ii) $\alpha = \beta = 1, \gamma = 0$.

10.2 Prove Theorem 10.3 (with $c = 10$) using the result of Karger and Stein [67] that in any weighted graph on n vertices the number of r-minimal cuts is $O\left((2n)^{2r}\right)$. (A cut $(S, V \setminus S)$, $S \subseteq V$, in a weighted graph G is called *r-minimal* if its weight, i.e., the sum of weights of the edges connecting S with $V \setminus S$, is at most r times the weight of the minimal weighted cut of G.)

10.3 Let X_k denote the number of vertices of degree k in a random graph $\mathbb{G}_{n,\mathbf{P^w}}$. Show that

$$\mathbb{E}\, X_k = \frac{1}{k!} \sum_{i=1}^{n} w_i e^{-w_i} + O\left(\frac{\tilde{w}\tilde{w}_{(k-2)}}{k!}\right),$$

where $\tilde{w} = \overline{w^2} = \rho \sum_{i=1}^{n} w_i^2$, $\rho = 1/W$, $W = \sum_{i=1}^{n} w_i$ and $\tilde{w}_{(t)} = \rho \sum_{i=1}^{n} w_i^t$.

10.4 Let X_k denote the number of vertices of degree k in a random graph $\mathbb{G}_{n,\mathbf{P^w}}$, where for $i = 1, 2, \ldots, n$ we have weights $w_i = ci^{-1/(\beta-1)}$ and $c, \beta > 0$ are constants. Show that

$$\mathbb{E}\, X_k \sim c' \frac{\Gamma(k - \beta + 1)}{\Gamma(k + 1)} \sim k^{-\beta},$$

for some constant c'. Here Γ is the Euler gamma function.

10.5 Let A be the adjacency matrix of $\mathbb{G}_{n,\mathbf{P^w}}$ and for a fixed value of x, let

$$c_i = \begin{cases} w_i & w_i > x, \\ x & w_i \leq x. \end{cases}$$

Let $m = \max\{w_i : i \in [n]\}$. Let $X_i = \frac{1}{c_i} \sum_{j=1}^{n} c_j a_{i,j}$. Show that

$$\mathbb{E}\, X_i \leq \overline{w^2} + x \text{ and } \operatorname{Var} X_i \leq \frac{m}{x}\overline{w^2} + x.$$

10.6 Suppose that $1 \leq w_i \ll W^{1/2}$ for $1 \leq i \leq n$ and that $w_i w_j \overline{w^2} \gg W \log n$. Show that w.h.p. diameter$(\mathbb{G}_{n,\mathbf{P^w}}) \leq 2$.

10.7 Show that w.h.p. a random 2-regular graph on n vertices consists of $O(\log n)$ vertex disjoint cycles.

10.8 Let H be a subgraph of $\mathbb{G}_{n,r}$, $r \geq 3$ obtained by independently including each vertex with probability $\frac{1+\varepsilon}{r-1}$, where $\varepsilon > 0$ is small and positive. Show that w.h.p. H contains a component of size $\Omega(n)$.

10.9 Extend the configuration model to bipartite graphs and show that a random r-regular bipartite graph is connected w.h.p. for $r \geq 3$.

10.10 Show that w.h.p. $\mathbb{G}_{n,r}$ is not planar for $r \geq 3$.

11 Small World

In an influential paper, Milgram [89] described the following experiment. He chose a person X to receive mail and then randomly chose a person Y to send it to. If Y did not know X, then Y was to send the mail to someone he/she thought more likely to know X and so on. Surprisingly, the mail got through in 64 out of 296 attempts, and the number of links in the chain was relatively small, between 5 and 6. Milgram's experiment suggests that large real-world networks although being globally sparse, in terms of the number of edges, have their nodes/vertices connected by relatively short paths. In addition, such networks are locally dense, i.e., vertices lying in a small neighborhood of a given vertex are connected by many edges. This observation is called the "small-world" phenomenon, and it has generated many attempts, both theoretical and experimental, to build and study appropriate models of small-world networks. The first attempt to explain this phenomenon and to build a more realistic model was introduced by Watts and Strogatz in 1998 in *Nature* (see [113]) followed by the publication of an alternative approach by Kleinberg in 2000 (see [72]). This chapter is devoted to the presentation of both models.

11.1 Watts–Strogatz Model

The Watts–Strogatz model starts with a kth power of an n-vertex cycle, denoted here as C_n^k. To construct it, fix n and k, $n \geq k \geq 1$ and take the vertex set as $V = [n] = \{1, 2, \ldots, n\}$ and edge-set $E = \{\{i, j\} : i + 1 \leq j \leq i + k\}$, where the additions are taken modulo n.

In particular, $C_n^1 = C_n$ is a cycle on n vertices. For an example of a square C_n^2 of C_n, see Figure 11.1.

Note that for $n > 2k$ graph C_n^k is $2k$-regular and has nk edges. Now choose each of the nk edges of C_n^k, one by one, and independently with small probability p decide to "rewire" it or leave it unchanged. The procedure goes as follows. We start, say, at vertex labeled 1, and move clockwise k times around the cycle. At the ith passage of the cycle, at each visited vertex, we take the edge connecting it to its neighbor at distance i to the right and decide, with probability p, if its other endpoint should be replaced by a uniformly random vertex of the cycle. However, we do not allow the creation of double

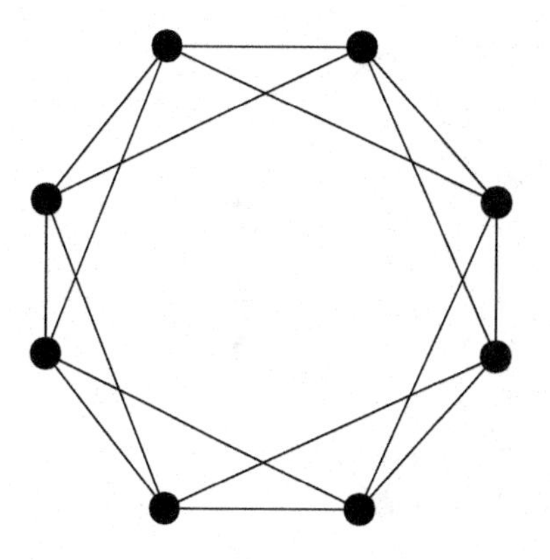
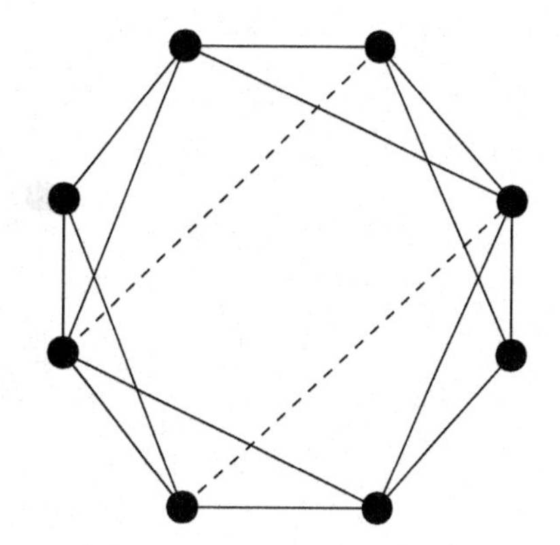

Figure 11.1 C_8^2 and C_8^2 after two rewirings

edges. Note that after this k round procedure is completed, the number of edges of the Watts–Strogatz random graph is kn, i.e., the same as in the "starting" graph C_n^k.

To study the properties of the original Watts–Strogatz model in a formal mathematical manner has proved rather difficult. Therefore, Newman and Watts (see [96]) proposed a modified version, where instead of rewiring the edges of C_n^k, each of the $\binom{n}{2} - nk$ edges not in C_n^k are added independently with probability p. Denote this random graph by $C_{n,p}^k$. In fact, $C_{n,p}^1$, when the starting graph is the n-vertex cycle, was introduced earlier by Ball, Mollison, and Scalia-Tomba [11] as "the great circle" epidemic model. For rigorous results on typical distances in $C_{n,p}^k$, see the papers of Barbour and Reinert [15, 16].

Gu and Huang [56] suggested a variant of the Newman–Watts random graph $C_{n,p}^k$ as a better approximation of the structure of the Watts–Strogatz random graph, where in addition to randomly sampling the edges not in C_n^k, each of the kn edges of C_n^k is removed independently with probability q (i.e., it remains in C_n^k with probability $1 - q$). Denote this random graph by $C_{n,p,1-q}^k$ and note that if $p = 1 - q$, then $C_{n,p,1-q}^k \equiv G_{n,1-q}$, while if we choose $q = 0$, it is equivalent to the Newman–Watts random graph $C_{n,p}^k$. Now, one can, for example, fix $q \neq 0$ and find $p = p(n, k, q)$ such that the expected number of edges of $C_{n,p,1-q}^k$ is equal to kn, i.e., is the same as the number of edges in the Watts and Strogatz random graph (see Exercise 11.1.1).

Much earlier Bollobás and Chung [23] took yet another approach to introducing "shortcuts" in C_n. Namely, let C_n be a cycle with n vertices labeled clockwise $1, 2, \ldots, n$, so that vertex i is adjacent to vertex $i + 1$ for $1 \leq i \leq n - 1$. Consider the graph G_n obtained by adding a randomly chosen perfect matching to C_n. (We will assume that n is even. For odd n one can add a random near-prefect matching.) Note that the graphs generated by this procedure are three-regular (see Figure 11.2).

It is easy to see that a cycle C_n itself has a diameter $n/2$. Bollobás and Chung proved that the diameter drops dramatically after adding to C_n such a system of random "shortcuts."

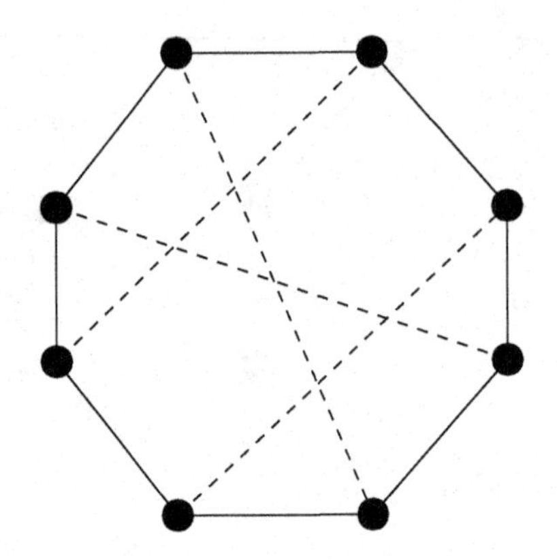

Figure 11.2 $C_8 \cup M$

Theorem 11.1 *Let G_n be formed by adding a random perfect matching M to an n-cycle C_n. Then w.h.p.*

$$\operatorname{diam}(G_n) \leq \log_2 n + \log_2 \log n + 10.$$

Proof For a vertex u of G_n define sets

$$S_i(u) = \{v : \operatorname{dist}(u, v) = i\} \quad \text{and} \quad S_{\leq i}(u) = \bigcup_{j \leq i} S_j(u),$$

where $\operatorname{dist}(u, v) = \operatorname{dist}_{G_n}(u, v)$ denotes the length of a shortest path between u and v in G_n.

Now define the following process for generating sets $S_i(u)$ and $S_{\leq i}(u)$ in G_n, Start with a fixed vertex u and "uncover" the chord (edge of M) incident to vertex u. This determines the set $S_1(u)$. Then we add the neighbors of $S_1(u)$ one by one to determine $S_2(u)$ and proceed to determine $S_i(u)$.

A chord incident to a vertex in $S_i(u)$ is called *inessential at level i* if the other vertex in $S_i(u)$ is within distance $3 \log_2 n$ in C_n of the vertices determined so far. Note that $|S_{\leq i}(u)| \leq 3 \cdot 2^i$ and so

$$\mathbb{P}(\text{a chord is inessential at level } i \mid S_{\leq i-1}(u)) \leq \frac{18 \cdot 2^{i+1} \log_2 n}{n}. \tag{11.1}$$

Denote by $\mathcal{A}$ the event that for every vertex u, at most one of the chords chosen in $S_{\leq i}(u)$ is inessential and suppose that $i \leq \frac{1}{5} \log_2 n$. Then

$$\begin{aligned}\mathbb{P}(\mathcal{A}^c) &= \mathbb{P}(\exists u : \text{at least two of the chords chosen in } S_{\leq i}(u) \text{ are inessential}) \\ &\leq n \binom{3 \cdot 2^{i+1}}{2} \left(\frac{18 \cdot 2^{i+1} \log_2 n}{n} \right)^2 = O\left(n^{-1/5} (\log n)^2\right).\end{aligned}$$

For a fixed vertex u, consider those vertices v in $S_i(u)$ for which there is a unique path from u to v of length i, say $u = u_0, u_1, \ldots, u_{i-1}, u_i = v$, such that

(i) if u_{i-1} is adjacent to v on the cycle C_n, then $S_{\leq i}(u)$ contains no vertex on C_n within distance $3 \log_2 n$ on the opposite side to v (denote the set of such vertices v by $C_i(u)$),

(ii) if $\{u_{i-1}, v\}$ is a chord, then $S_{\leq i}(u) \setminus \{v\}$ contains no vertex within distance $3 \log_2 n$ both to the left and to the right of v (denote the set of such vertices by $D_i(u)$).

Obviously,

$$C_i(u) \cup D_i(u) \subseteq S_i(u).$$

Note that if the event $\mathcal{A}$ holds, then, for $i \leq \frac{1}{5} \log_2 n$,

$$|C_i(u)| \geq 2^{i-2} \quad \text{and} \quad |D_i(u)| \geq 2^{i-3}. \tag{11.2}$$

Let $\frac{1}{5} \log_2 n \leq i \leq \frac{3}{5} \log_2 n$. Denote by $\mathcal{B}$ the event that for every vertex u, at most $2^i n^{-1/10}$ inessential chords leave $S_i(u)$. There are at most 2^i chords leaving $S_i(u)$ for such is and so by (11.1), for large n,

$$\begin{aligned}\mathbb{P}(\mathcal{B}^c) &= \mathbb{P}(\exists u : \text{ at least } 2^i n^{-1/10} \text{ inessential chords leave } S_i(u)) \\ &\leq n \binom{2^i}{2^i\ n^{-1/10}} = O\left(n^{-2}\right). \end{aligned} \tag{11.3}$$

For $v \in C_i(u)$, a new neighbor of v in C_n is a potential element of $C_{i+1}(u)$ and a new neighbor, which is the end-vertex of the chord from v, is a potential element of $D_{i+1}(u)$. Also, if $v \in D_i(u)$, then the two neighbors of v in C_n are potential elements of $C_{i+1}(u)$. Here "potential" means that the vertices in question become elements of $C_{i+1}(u)$ and $D_{i+1}(u)$ unless the corresponding edge is inessential.

Assuming that the events $\mathcal{A}$ and $\mathcal{B}$ both hold and $\frac{1}{5} \log_2 n \leq i \leq \frac{3}{5} \log_2 n$, then

$$|C_{i+1}(u)| \geq |C_i(u)| + 2|D_i(u)| - 2^{i+1} n^{-1/10},$$
$$|D_{i+1}(u)| \geq |C_i(u)| - 2^{i+1} n^{-1/10},$$

while for $i \leq \frac{1}{5} \log_2 n$ the bounds given in (11.2) hold. Hence, for all $3 \leq i \leq \frac{3}{5} \log_2 n$, we have

$$|C_i(u)| \geq 2^{i-3} \quad \text{and} \quad |D_i(u)| \geq 2^{i-4}.$$

To finish the proof set,

$$i_0 = \left\lceil \frac{\log_2 n + \log_2 \log n + c}{2} \right\rceil,$$

where $c \geq 9$ is a constant.

Let us choose chords leaving $C_{i_0}(u)$ one by one. At each choice, the probability of not selecting the other end-vertex in $C_{i_0}(u)$ is at most $1 - (2^{i_0-3}/n)$. Since we have to make at least $|C_{i_0}(u)|/2 \geq 2^{i_0-4}$ such choices, we have

$$\mathbb{P}(\text{dist}(u, v) > 2i_0 + 1 | \mathcal{A} \cap \mathcal{B}) \leq \left(1 - \frac{2^{i_0-3}}{n}\right)^{2^{i_0-4}} \leq n^{-4}. \tag{11.4}$$

Hence,

$$\mathbb{P}(\text{diam}(G_n) > 2i_0 + 1) \leq \mathbb{P}(\mathcal{A}^c) + \mathbb{P}(\mathcal{B}^c) + \sum_{u,v} \mathbb{P}(\text{dist}(u, v) > 2i_0 + 1 | \mathcal{A} \cap \mathcal{B})$$

$$\leq c_1(n^{-1/5}(\log n)^2) + c_2 n^{-2} + n^{-2} = o(1).$$

Therefore, w.h.p. the random graph G_n has diameter at most

$$2\left\lceil \frac{\log_2 n + \log_2 n \log n + 9}{2} \right\rceil \leq \log_2 n + \log_2 n \log n + 10,$$

which completes the proof of Theorem 11.1. □

In fact, based on the similarity between a random three-regular graph and the graph G_n defined previously, one can prove more precise bounds, showing (see Wormald [115]) that w.h.p. $\mathrm{diam}(G_n)$ is highly concentrated, i.e., that

$$\log_2 n + \log_2 n \log n - 4 \leq \mathrm{diam}(G_n) \leq \log_2 n + \log_2 n \log n + 4.$$

The above-mentioned result illustrates one of the two main characteristics of the small-world phenomenon. Namely, it shows that although real-world networks are sparse, the distance between any two nodes is small. The second typical property of networks is that they are at the same time "locally dense" or, alternatively, have "a high degree of clustering." To illustrate this notion, let us formally introduce definitions of local and global clustering for graphs. Let G be a graph on the vertex set $[n]$. Let v be a vertex in G with degree $\deg(v) = d(v)$. Let $N(v)$ denote the neighborhood of v, i.e., the set of all vertices incident to v and $e_{N(v)}$ be the number of edges in the subgraph of G induced by the vertices in $N(v)$. Note that $|N(v)| = d(v)$.

Define the *local clustering coefficient* of v as

$$lcc(v) = \frac{e_{N(v)}}{\binom{d(v)}{2}}. \tag{11.5}$$

Hence, it is natural to introduce *the average local clustering coefficient* as a measure of the local density of a graph G, so

$$C_1(G) = \frac{1}{n} \sum_{v=1}^{n} lcc(v). \tag{11.6}$$

Another way of averaging the number of edges in the neighborhood of each vertex might be more informative in measuring the total clustering of a graph G. Let

$$C_2(G) = \frac{\sum_{v=1}^{n} e_{N(v)}}{\sum_{v=1}^{n} \binom{d(v)}{2}}. \tag{11.7}$$

In the next lemma, we separately find the expected values of the two basic ingredients of both clustering coefficients $C_1(G)$ and $C_2(G)$, namely $\mathbb{E}(e_{N(v)})$ and $\mathbb{E}\left(\binom{d(v)}{2}\right)$, for any vertex v of the random graph $C^k_{n,p,1-q}$, i.e, of the generalized Newman–Watts graph, introduced by Gu and Huan.

Lemma 11.2 *Let $G = C^k_{n,p,1-q}$. Then for every vertex v,*

$$\mathbb{E}\left(\binom{d(v)}{2}\right) =$$
$$\frac{1}{2}\left\{2k(2k-1)(1-q)^2 + (n-2k)(n-2k-1)p^2\right\} + 2k(1-q)(n-2k)p. \quad (11.8)$$

$$\mathbb{E}(e_{N(v)}) =$$
$$(1+2(k-1))(1-q)^3 + (2k-1)(n-2k-1)p^2(1-q) + \binom{n-2k-1}{2}p^3. \quad (11.9)$$

Proof
To prove the first statement, we use the fact that if Z is distributed as the binomial $Bin(n,p)$, then

$$\mathbb{E}\left(\binom{Z}{2}\right) = \frac{1}{2}n(n-1)p^2. \quad (11.10)$$

The number of edges of C_n^k incident to v that are not deleted is distributed as $Bin(2k, 1-q)$, and this explains the first term in (11.8). The number of added edges incident to v is distributed as $Bin(n-2k, p)$, and this explains the second term in (11.8). The quantities giving rise to the first two terms are independent and so the third term is just the product of their expectations.

To show that (11.9) holds, note that counting edges in the neighborhood of v is the same as counting triangles in the subgraph of $C_{n,p,1-q}^k$ induced by $v \cup N(v)$. Note that there are $1+2(k-1)$ triangles in C_n^k and each exists with probability $(1-q)^3$. There are $(2k-1)(n-2k-1)$ triangles that use one edge of C_n^k and each exists with probability $p^2(1-q)$. There are $\binom{n-2k-1}{2}$ triangles that do not use any edge of C_n^k and each exists with probability p^3. This proves (11.9). □

Following the observation in the proof of (11.9), one can immediately see that in fact,

$$C_2(G) = \frac{6T}{\sum_{v=1}^{n} d(v)(d(v)-1)}, \quad (11.11)$$

where $T = T(G)$ is the number of triangles in a graph G. For $G = C_{n,p,1-q}^k$, the expectation $\mathbb{E}(T) = n\,\mathbb{E}(e_{N(v)})$, where $\mathbb{E}(e_{N(v)})$ is given by (11.9).

Exercises

11.1.1 Fix $q \neq 0$ and find $p = p(n,k,q)$ such that the expected number of edges of a random graph $C_{n,p,1-q}^k$ is equal to kn, i.e., is the same as the number of edges in the Watts–Strogatz random graph.
11.1.2 Verify equation (11.3).
11.1.3 Verify equation (11.4).
11.1.4 Find the value $lcc(v)$ for any vertex v of C_n^k, the kth power of a cycle C_n.
11.1.5 Assume that p, q are constants in $C_{n,p,1-q}$. Use McDiarmid's inequality (see Lemma 9.6) to show that the quantities in Lemma 11.2 are both concentrated around their means. (Here take the $W_i, i = 1, \ldots, \binom{n}{2}$ to be the indicator variables for the edges of K_n that exist in $C_{n,p,1-q}$.)

11.2 Kleinberg's Model

The model introduced by Kleinberg [72] can be generalized significantly, but to be specific we consider the following. We start with the $n \times n$ grid G_0 which has vertex set $[n]^2$ and where (i, j) is adjacent to (i', j') if and only if $d((i, j), (i', j')) = 1$, where $d((i, j), (k, \ell)) = |i - k| + |j - \ell|$. In addition, each vertex $u = (i, j)$ will choose another random neighbor $\varphi(u)$, where

$$\mathbb{P}(\varphi(u) = v = (k, \ell)) = \frac{d(u, v)^{-2}}{D_u} \tag{11.12}$$

and

$$D_x = \sum_{y \neq x} d(x, y)^{-2}.$$

The random neighbors model "long-range contacts." Let the grid G_0 plus the extra random edges be denoted by G.

It is not difficult to show that w.h.p. these random contacts reduce the diameter of G to order $\log n$. This, however, would not explain Milgram's success. Instead, Kleinberg proposed the following decentralized algorithm $\mathcal{A}$ for finding a path from an initial vertex $u_0 = (i_0, j_0)$ to a target vertex $u_\tau = (i_\tau, j_\tau)$: when at u move to the neighbor closest in distance to u_τ.

Theorem 11.3 *Algorithm $\mathcal{A}$ finds a path from initial to target vertex of order $O((\log n)^2)$, in expectation.*

Proof Note that each step of $\mathcal{A}$ finds a node closer to the target than the current node and so the algorithm must terminate with a path.

Observe next that for any vertex x of G we have

$$D_x \leq \sum_{j=1}^{2n-2} 4j \times j^{-2} = 4 \sum_{j=1}^{2n-2} j^{-1} \leq 4 \log(3n).$$

As a consequence, v is the long-range contact of vertex u, with probability at least $(4 \log(3n) d(u, v)^2)^{-1}$.

For $0 < j \leq \log_2 n$, we say that the execution of $\mathcal{A}$ is in Phase j if the distance of the current vertex u to the target is greater than 2^j, but at most 2^{j+1}. We say that $\mathcal{A}$ is in Phase 0 if the distance from u to the target is at most 2.

Let B_j denote the set of nodes at distance 2^j or less from the target. Then

$$|B_j| > 2^{2j-1}. \tag{11.13}$$

Note that by the triangle inequality, each member of B_j is within distance $2^{j+1} + 2^j < 2^{2j+2}$ of u.

Let $X_j \leq 2^{j+1}$ be the time spent in Phase j. Assume first that

$\log_2 \log_2 n \leq j \leq \log_2 n$. Phase j will end if the long-range contact of the current vertex lies in B_j. The probability of this is at least

$$\frac{2^{2j-1}}{4\log(3n)2^{2j+4}} = \frac{1}{128\log(3n)}.$$

We can reveal the long-range contacts as the algorithm progresses. In this way, the long-range contact of the current vertex will be independent of the previous contacts of the path. Thus,

$$\mathbb{E} X_j = \sum_{i=1}^{\infty} \mathbb{P}(X_j \geq i) \leq \sum_{i=1}^{\infty} \left(1 - \frac{1}{128\log(3n)}\right)^i < 128\log(3n).$$

Now if $0 \leq j \leq \log_2 \log_2 n$, then $X_j \leq 2^{j+1} \leq 2\log_2 n$. Thus, the expected length of the path found by $\mathcal{A}$ is at most $128\log(3n) \times \log_2 n$. □

In the same paper, Kleinberg showed that replacing $d(u,v)^{-2}$ by $d(u,v)^{-r}$ for $r \neq 2$ led to non-polylogarithmic path length.

Exercises

11.2.1 Verify equation (11.13).

11.2.2 Let H be a fixed graph with minimum degree αn, where $0 < \alpha < 1$ is constant. Let R be a random set of $2\alpha^{-2}\log n$ random edges. Show that the graph $G = H + R$ has diameter two w.h.p.

Problems for Chapter 11

11.1 Let p be chosen so that the expected number of edges of the random graph $C^k_{n,p,1-q}$ is equal to kn. Prove that if $k = c\log n$, where $c < 1$ is a constant, then for a fixed $0 < q < 1$, the random graph $C^k_{n,p,1-q}$ is disconnected w.h.p.

11.2 Define a randomized community-based small-world random graph Γ_n as follows. Partition a set of n nodes into m communities (clusters), each of size k, i.e., $n = m \cdot k$ and let cluster i be a complete graph $K^{(i)}_k$, where $i = 1, 2, \ldots, m$. Next, represent each cluster as a vertex of an auxiliary binomial random graph $G_{m,p}$. If vertices i and j are connected in $G_{m,p}$ by an edge, we randomly pick a vertex in either $K^{(i)}_k$ or $K^{(j)}_k$ and link it to all vertices of the other cluster. After completing this task for all edges of $G_{m,p}$, we connect vertices of Γ_n which are linked to the same cluster.

Let $k \approx \ln n$ and $p = c\frac{\ln n}{n}$, where $c > 1$ is a constant. Bound the expected number of edges of Γ_n by $O(n\ln n)$ (see Cont and Tanimura [33]).

11.3 Use a similar argument as in the proof of Theorem 11.1 to show that if T is a complete binary tree on $2^k - 1$ vertices and we add two random matchings of size 2^{k-1} to the leaves of T, then the diameter of the resulting graph G satisfies $\text{diam}(G) \leq \log_2 n + \log_2 \log n + 10$ w.h.p. (see [23]).

11.4 Let $0 \leq r < 2$. Suppose now that we replace (11.12) by

$$\mathbb{P}(\varphi(u) = v = (k,\ell)) = \frac{d(u,v)^{-r}}{D_u}, \text{ where } D_x = \sum_{y \neq x} d(x,y)^{-r}.$$

Show that now the expected length of the path found by $\mathcal{A}$ of Section 11.2 is $\Omega(n^{(2-r)/3})$.

11.5 Suppose that in the previous problem we have $r > 2$. Show now that the expected length of the path found by $\mathcal{A}$ of Section 11.2 is $\Omega(n^{(r-2)/(r-1)})$.

11.6 Let H be an arbitrary connected graph of bounded degree Δ. Let R be a random set of cn random edges. Show that the graph $G = H + R$ has diameter $O(\log n)$ w.h.p. for sufficiently large $c > 0$.

12 Network Processes

Until now, we have considered "static" (in terms of the number of vertices) models of real-world networks only. However, more often the networks are constructed by some random "dynamic" process of adding vertices, together with some new edges connecting those vertices with the already existing network. Typically, networks grow (or shrink) during time intervals. It is a complex process, and we do not fully understand the mechanisms of their construction as well as the properties they acquire. Therefore, to model such networks is quite challenging and needs specific models of random graphs, possessing properties observed in a real-world network. One such property is that often the degree sequence exhibits a tail that decays polynomially, as opposed to classical random graphs, whose tails decay (super)exponentially (see, for example, Faloutsos, Faloutsos, and Faloutsos [47]). Grasping this property led to the development of so-called preferential attachment models, whose introduction is attributed to Barabási and Albert, but, in principle, known and studied earlier by random graph theorists as random plane-oriented recursive trees (see Chapter 14 of [52] for details and references). The Barabási–Albert [13] model is ambiguously defined, as pointed out by Bollobás, Riordan, Spencer, and Tusnády [28] (see also Bollobás, Riordan [27]). Section 12.1 studies a similar model to that of [28]. After the presentation of basic properties of the preferential attachment model, we conclude the first section with a brief discussion of its application to study the spread of infection through a network, called bootstrap percolation.

The last section of this chapter is devoted to a generalization of the preferential attachment model, called spatial preferential attachment. It combines simple preferential attachment with geometry by introducing "spheres of influence" of vertices, whose volumes depend on their in-degrees.

12.1 Preferential Attachment

Fix an integer $m > 0$, constant and define a sequence of graphs $G_1, G_2, \ldots, G_t$. The graph G_t has a vertex set $[t]$ and G_1 consists of m loops on vertex 1. Suppose we have constructed G_t. To obtain G_{t+1} we apply the following rule. We add vertex $t + 1$ and connect it to m randomly chosen vertices $y_1, y_2, \ldots, y_m \in [t]$ in such a way that for

$i = 1, 2, \ldots, m$,

$$\mathbb{P}(y_i = w) = \frac{\deg(w, G_t)}{2mt},$$

where $\deg(w, G_t)$ is the degree of w in G_t.

In this way, G_{t+1} is obtained from G_t by adding vertex $t+1$ and m randomly chosen edges, in such a way that the neighbors of $t+1$ are biased toward higher-degree vertices.

When $m = 1$, G_t is a tree and this is basically a plane-oriented recursive tree.

Expected Degree Sequence: Power Law

Fix t and let $V_k(t)$ denote the set of vertices of degree k in G_t, where $m \le k = \tilde{O}(t^{1/2})$. Let $D_k(t) = |V_k(t)|$ and $\bar{D}_k(t) = \mathbb{E}(D_k(t))$. Then

$$\mathbb{E}(D_k(t+1)|G_t) = D_k(t) + m\left(\frac{(k-1)D_{k-1}(t)}{2mt} - \frac{kD_k(t)}{2mt}\right) + 1_{k=m} + \varepsilon(k,t). \quad (12.1)$$

Explanation of (12.1)**:** The total degree of G_t is $2mt$ and so $\frac{(k-1)D_{k-1}(t)}{2mt}$ is the probability that y_i is a vertex of degree $k-1$, creating a new vertex of degree k. Similarly, $\frac{kD_k(t)}{2mt}$ is the probability that y_i is a vertex of degree k, destroying a vertex of degree k. At this point $t+1$ has degree m and this accounts for the term $1_{k=m}$. The term $\varepsilon(k,t)$ is an error term that accounts for the possibility that $y_i = y_j$ for some $i \neq j$.

Thus

$$\varepsilon(k,t) = O\left(\binom{m}{2}\frac{k}{mt}\right) = \tilde{O}(t^{-1/2}). \quad (12.2)$$

Taking expectations over G_t, we obtain

$$\bar{D}_k(t+1) = \bar{D}_k(t) + 1_{k=m} + m\left(\frac{(k-1)\bar{D}_{k-1}(t)}{2mt} - \frac{k\bar{D}_k(t)}{2mt}\right) + \varepsilon(k,t). \quad (12.3)$$

Under the assumption $\bar{D}_k(t) \sim d_k t$ (justified below), we are led to consider the recurrence

$$d_k = \begin{cases} 1_{k=m} + \frac{(k-1)d_{k-1} - kd_k}{2} & \text{if } k \ge m, \\ 0 & \text{if } k < m, \end{cases} \quad (12.4)$$

or

$$d_k = \begin{cases} \frac{k-1}{k+2}d_{k-1} + \frac{2\cdot 1_{k=m}}{k+2} & \text{if } k \ge m, \\ 0 & \text{if } k < m. \end{cases}$$

Therefore,

$$d_m = \frac{2}{m+2},$$

$$d_k = d_m \prod_{l=m+1}^{k} \frac{l-1}{l+2} = \frac{2m(m+1)}{k(k+1)(k+2)}. \tag{12.5}$$

So for large k, under our assumption $\bar{D}_k(t) \sim d_k t$, we see that

$$\bar{D}_k(t) \sim \frac{2m(m+1)}{k^3} t. \tag{12.6}$$

So the number of vertices of degree k decays like t/k^3. We refer to this as a *power law*, where the decay is polynomial in k, rather than (super)exponential as in $\mathbb{G}_{n,m}$, $m = cn$. We now show that the assumption $\bar{D}_k(t) \sim d_k t$ can be justified. Note that the following theorem is vacuous for $k \gg t^{1/6}$.

Theorem 12.1

$$|\bar{D}_k(t) - d_k t| = \tilde{O}(t^{1/2}) \textit{ for } k = \tilde{O}(t^{1/2}).$$

Proof Let

$$\Delta_k(t) = \bar{D}_k(t) - d_k t.$$

Then, replacing $\bar{D}_k(t)$ by $\Delta_k(t) + d_k t$ in (12.3) and using (12.2) and (12.4), we get

$$\Delta_k(t+1) = \frac{k-1}{2t}\Delta_{k-1}(t) + \left(1 - \frac{k}{2t}\right)\Delta_k(t) + \tilde{O}(t^{-1/2}). \tag{12.7}$$

Now assume inductively on t that for every $k \geq 0$,

$$|\Delta_k(t)| \leq At^{1/2}(\log t)^\beta,$$

where $(\log t)^\beta$ is the hidden power of logarithm in $\tilde{O}(t^{-1/2})$ of (12.7) and A is an unspecified constant.

This is trivially true for $k < m$ also for small t if we make A large enough. So, replacing $\tilde{O}(t^{-1/2})$ in (12.7) by the more explicit $\alpha t^{-1/2}(\log t)^\beta$, we get

$$\begin{aligned}\Delta_k(t+1) &\leq \left|\frac{k-1}{2t}\Delta_{k-1}(t)\right| + \left|\left(1 - \frac{k}{2t}\right)\Delta_k(t)\right| + \alpha t^{-1/2}(\log t)^\beta \\ &\leq (\log t)^\beta (At^{1/2} + \alpha t^{-1/2}).\end{aligned} \tag{12.8}$$

Note that if t is sufficiently large, then

$$(t+1)^{1/2} = t^{1/2}\left(1 + \frac{1}{t}\right)^{1/2} \geq t^{1/2} + \frac{1}{3t^{1/2}},$$

and so

$$\begin{aligned}\Delta_k(t+1) &\leq (\log(t+1))^\beta \left(A\left[(t+1)^{1/2} - \frac{1}{3t^{1/2}}\right] + \frac{\alpha}{t^{1/2}}\right) \\ &\leq A\,(\log(t+1))^\beta\,(t+1)^{1/2},\end{aligned}$$

assuming that $A \geq 3\alpha$. □

In the next section, we will justify our bound of $\tilde{O}(t^{1/2})$ for vertex degrees. After that we will prove the concentration of the number of vertices of degree k for small k.

Maximum Degree

Fix $s \leq t$ and let X_l be the degree of vertex s in G_l for $s \leq l \leq t$. We prove the following high-probability upper bound on the degree of vertex s.

Lemma 12.2

$$\mathbb{P}(X_t \geq Ae^m (t/s)^{1/2} (\log(t+1))^2) = O(t^{-A}).$$

Proof Note first that $X_s = m$. If $0 < \lambda < \varepsilon_t = \frac{1}{\log(t+1)}$, then

$$\begin{aligned}
&\mathbb{E}\left(e^{\lambda X_{l+1}} | X_l\right) \\
&= e^{\lambda X_l} \sum_{k=0}^{m} \binom{m}{k} \left(\frac{X_l}{2ml}\right)^k \left(1 - \frac{X_l}{2ml}\right)^{m-k} e^{\lambda k} \\
&\leq e^{\lambda X_l} \sum_{k=0}^{m} \binom{m}{k} \left(\frac{X_l}{2ml}\right)^k \left(1 - \frac{X_l}{2ml}\right)^{m-k} (1 + k\lambda(1 + k\lambda)) \\
&= e^{\lambda X_l} \left(1 + \frac{\lambda(1+\lambda) X_l}{2l} + \frac{(m-1)\lambda^2 X_l^2}{4ml^2}\right) \\
&\leq e^{\lambda X_l} \left(1 + \frac{\lambda X_l}{2l}(1 + m\lambda)\right), \qquad \text{since } X_l \leq 2ml, \\
&\leq e^{\lambda\left(1 + \frac{(1+m\lambda)}{2l}\right) X_l}.
\end{aligned}$$

We define a sequence $\lambda = (\lambda_s, \lambda_{s+1}, \ldots, \lambda_t)$ where

$$\lambda_{j+1} = \left(1 + \frac{1 + m\lambda_j}{2j}\right) \lambda_j < \varepsilon_t.$$

Here our only choice will be λ_s. We show below that we can find a suitable value for this, but first observe that if we manage this then

$$\mathbb{E}\left(e^{\lambda_s X_t}\right) \leq \mathbb{E}\left(e^{\lambda_{s+1} X_{t-1}}\right) \cdots \leq \mathbb{E}\left(e^{\lambda_t X_s}\right) \leq 1 + o(1).$$

Now

$$\lambda_{j+1} \leq \left(1 + \frac{1 + m\varepsilon_t}{2j}\right) \lambda_j$$

implies that

$$\lambda_t = \lambda_s \prod_{j=s}^{t} \left(1 + \frac{1 + m\varepsilon_t}{2j}\right) \leq \lambda_s \exp\left\{\sum_{j=s}^{t} \frac{1 + m\varepsilon_t}{2j}\right\} \leq e^m \left(\frac{t}{s}\right)^{1/2} \lambda_s.$$

So a suitable choice for $\lambda = \lambda_s$ is

$$\lambda_s = e^{-m}\varepsilon_t\left(\frac{s}{t}\right)^{1/2}.$$

This gives

$$\mathbb{E}\left(\exp\left\{e^{-m}\varepsilon_t(s/t)^{1/2}X_t\right\}\right) \leq 1 + o(1).$$

So,

$$\mathbb{P}\left(X_t \geq Ae^m(t/s)^{1/2}(\log(t+1))^2)\right) \leq$$
$$e^{-\lambda_t Ae^m(t/s)^{1/2}(\log(t+1)^2)}\,\mathbb{E}\left(e^{\lambda_s X_t}\right) = O(t^{-A})$$

by the Markov inequality. □

Thus with probability $(1 - o(1))$ as $t \to \infty$, we have that the maximum degree in G_t is $O(t^{1/2}(\log t)^2)$. This is not the best possible. One can prove that w.h.p. the maximum degree is $O(t^{1/2}\omega(t))$ and $\Omega(t^{1/2}/\omega(t))$ for any $\omega(t) \to \infty$, see, for example, Flaxman, Frieze, and Fenner [49].

Concentration of Degree Sequence

Fix a value k for a vertex degree. We show that $D_k(t)$ is concentrated around its mean $\bar{D}_k(t)$.

Theorem 12.3

$$\mathbb{P}(|D_k(t) - \bar{D}_k(t)| \geq u) \leq 2\exp\left\{-\frac{u^2}{32mt}\right\}. \tag{12.9}$$

Proof Let $Y_1, Y_2, \ldots, Y_{mt}$ be the sequence of edge choices made in the construction of G_t, and for $Y_1, Y_2, \ldots, Y_i$, let

$$Z_i = Z_i(Y_1, Y_2, \ldots, Y_i) = \mathbb{E}(D_k(t) \mid Y_1, Y_2, \ldots, Y_i). \tag{12.10}$$

Note that the sequence $Z_0, Z_1, \ldots, Z_{mt}$, where $Z_0 = \mathbb{E}(D_k(t))$, is a martingale. We will prove next that $|Z_i - Z_{i-1}| \leq 4$ and then (12.9) follows directly from the Azuma–Hoeffding inequality, see Lemma 10.14.

Fix $Y_1, Y_2, \ldots, Y_i$ and $\hat{Y}_i \neq Y_i$. We define a map (measure preserving projection) φ of

$$Y_1, Y_2, \ldots, Y_{i-1}, Y_i, Y_{i+1}, \ldots, Y_{mt}$$

to

$$Y_1, Y_2, \ldots, Y_{i-1}, \hat{Y}_i, \hat{Y}_{i+1}, \ldots, \hat{Y}_{mt}$$

such that

$$|Z_i(Y_1, Y_2, \ldots, Y_i) - Z_i(Y_1, Y_2, \ldots, \hat{Y}_i)| \leq 4. \tag{12.11}$$

In the preferential attachment model, we can view vertex choices in the graph G as

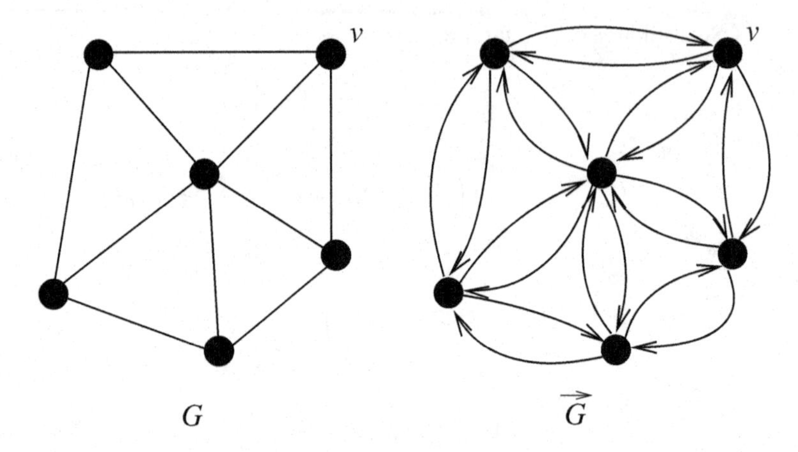

Figure 12.1 Constructing $\vec{G}$ from G

random choices of arcs in a digraph $\vec{G}$, which is obtained by replacing every edge of G by a directed two-cycle (see Figure 12.1).

Indeed, if we choose a random arc and choose its head, then v will be chosen with probability proportional to the number of arcs with v as the head, i.e., its degree. Hence $Y_1, Y_2, \dots$ can be viewed as a sequence of arc choices. Let

$$Y_i = (x, y) \text{ where } x > y, \tag{12.12}$$

$$\hat{Y}_i = (\hat{x}, \hat{y}) \text{ where } \hat{x} > \hat{y}. \tag{12.13}$$

Note that $x = \hat{x}$ if $i \bmod m \neq 1$.

Now suppose $j > i$ and $Y_j = (u, v)$ arises from choosing (w, v). Then we define

$$\varphi(Y_j) = \begin{cases} Y_j & (w, v) \neq Y_i, \\ (w, \hat{y}) & (w, v) = Y_i. \end{cases} \tag{12.14}$$

This map is measure preserving since each sequence $\varphi(Y_1, Y_2, \dots, Y_t)$ occurs with the probability $\prod_{j=i+1}^{tm} j^{-1}$. Only $x, \hat{x}, y, \hat{y}$ change the degree under the map φ so $D_k(t)$ changes by at most four. □

We will now study the degrees of early vertices.

Degrees of Early Vertices

Let $d_t(s)$ denote the degree of vertex s at time t. Then, we have $d_s(s) = m$ and

$$\mathbb{E}(d_{t+1}(s)|G_t) = d_t(s) + \frac{md_t(s)}{2mt} = d_t(s)\left(1 + \frac{1}{2t}\right).$$

So, because

$$\frac{2^{2s+1}s!(s-1)!}{(2s)!} \sim 2\left(\frac{\pi}{s}\right)^{1/2} \text{ for large } s,$$

we have

$$\mathbb{E}(d_s(t)) \sim m\left(\frac{t}{s}\right)^{1/2} \text{ for large } s. \tag{12.15}$$

For random variables X, Y and a sequence of random variables $\mathcal{Z} = Z_1, Z_2, \ldots, Z_k$ taking discrete values, we write (as in (2.37))

$$X \succ Y \text{ to mean that } \mathbb{P}(X \geq a) \geq \mathbb{P}(Y \geq a),$$

and

$$X \mid_{\mathcal{Z}} \succ Y \mid_{\mathcal{Z}} \text{ to mean that}$$
$$\mathbb{P}(X \geq a \mid Z_l = z_l, l = 1, \ldots, k) \geq \mathbb{P}(Y \geq a \mid Z_l = z_l, l = 1, \ldots, k)$$

for all choices of a, z.

Fix $i \leq j-2$ and let $X = d_{j-1}(t), Y = d_j(t)$ and $Z_l = d_l(t), l = i, \ldots, j-2$.

Lemma 12.4 $X \mid_{\mathcal{Z}} \succ Y \mid_{\mathcal{Z}}$.

Proof Consider the construction of $G_{j+1}, G_{j+2}, \ldots, G_t$. We condition on those edge choices of $j+1, j+2, \ldots, t$ that have one end in $i, i+1, \ldots, j-2$. Now if vertex j does not choose an edge $(j-1, j)$ then the conditional distributions of $d_{j-1}(t), d_j(t)$ are identical. If vertex j does choose edge $(j-1, j)$ and we do not include this edge in the value of the degree of $j-1$ at times $j+1$ onwards, then the conditional distributions of $d_{j-1}(t), d_j(t)$ are again identical. Ignoring this edge will only reduce the chances of $j-1$ being selected at any stage and the lemma follows. □

Corollary 12.5 *If $j \geq i-2$, then $d_i(t) \succ (d_{i+1}(t) + \cdots + d_j(t))/(j-i)$.*

Proof Fix $i \leq l \leq j$ and then we argue by induction that

$$d_{i+1}(t) + \cdots + d_l(t) + (j-l)d_{l+1}(t) \prec d_{i+1}(t) + \cdots + (j-l+1)d_l(t). \tag{12.16}$$

This is trivial for $j = l$ as the LHS is then the same as the RHS. Also, if true for $l = i$, then

$$d_{i+1}(t) + \cdots + d_j(t) \prec (j-i)d_{i+1}(t) \prec (j-i)d_i(t)$$

where the second inequality follows from Lemma 12.4 with $j = i+1$.

Putting $\mathcal{Z} = d_{i+1}(t), \ldots, d_{l-1}(t)$, we see that (12.16) is implied by

$$d_l(t) + (j-l)d_{l+1}(t) \mid_{\mathcal{Z}} \prec (j-l+1)d_l(t) \mid_{\mathcal{Z}} \text{ or } d_{l+1}(t) \mid_{\mathcal{Z}} \prec d_l(t) \mid_{\mathcal{Z}}$$

after subtracting $(j-l)d_{l+1}(t)$. But the latter follows from Lemma 12.4. □

Lemma 12.6 *Fix $1 \leq s = O(1)$ and let $\omega = \log^2 t$ and let $D_s(t) = \sum_{i=s+1}^{s+\omega} d_s(t)$. Then w.h.p. $D_s(t) \sim 2m(\omega t)^{1/2}$.*

Proof

We have from (12.15) that

$$\mathbb{E}(D_s(t)) \sim m \sum_{i=s+1}^{s+\omega} \left(\frac{t}{i}\right)^{1/2} \sim 2m(\omega t)^{1/2}.$$

Going back to the proof of Theorem 12.3 we consider the map φ as defined in (12.14). Unfortunately, (12.11) does not hold here. But we can replace 4 by $10\log t$, *most of the time*. So we let $Y_1, Y_2, \ldots, Y_{mt}$ be as in Theorem 12.3. Then let ψ_i denote the number of times that $(w, v) = Y_i$ in equation (12.14). Now ψ_j is the sum of $mt - j$ independent Bernoulli random variables and $\mathbb{E}(\psi_i) \leq \sum_{j=i+1}^{mt} 1/mj \leq m^{-1}\log mt$. It follows from the Chernoff–Hoeffding inequality that $\mathbb{P}(\psi_i \geq 10\log t) \leq t^{-10}$. Given this, we define a new random variable $\widehat{d}_s(t)$ and let $\widehat{D}_s(t) = \sum_{j=1}^{\omega} \widehat{d}_{s+j}(t)$. Here $\widehat{d}_{s+j}(t) = d_{s+j}(t)$ for $j = 1, 2, \ldots, \omega$ unless there exists i such that $\psi_i \geq 10\log t$. If there is an i such that $\psi_i \geq 10\log t$, then assuming that i is the first such we let $\widehat{D}_s(t) = Z_i(Y_1, Y_2, \ldots, Y_i)$ where Z_i is as defined in (12.10), with $D_k(t)$ replaced by $\widehat{D}_s(t)$. In summary, we have

$$\mathbb{P}(\widehat{D}_s(t) \neq D_s(t)) \leq t^{-10}. \tag{12.17}$$

So,

$$|\,\mathbb{E}(\widehat{D}_s(t)) - \mathbb{E}(D_s(t)| \leq t^{-9}.$$

And finally,

$$|Z_i - Z_{i-1}| \leq 20\log t.$$

This is because each $Y_i, \hat{Y}_i$ concerns at most two of the vertices $s+1, s+2, \ldots, s+\omega$. So,

$$\mathbb{P}(|\widehat{D}_s(t) - \mathbb{E}(\widehat{D}_s(t))| \geq u) \leq \exp\left\{-\frac{u^2}{800mt\log^2 t}\right\}. \tag{12.18}$$

Putting $u = \omega^{3/4}t^{1/2}$ into (12.18) yields the claim. □

Combining Corollary 12.5 and Lemma 12.6, we have the following theorem.

Theorem 12.7 *Fix $1 \leq s = O(1)$ and let $\omega = \log^2 t$. Then w.h.p. $d_i(t) \geq mt^{1/2}/\omega^{1/2}$ for $i = 1, 2, \ldots, s$.*

Proof Corollary 12.5 and (12.17) imply that $d_i(t) > D_i(t)/\omega$. Now apply Lemma 12.6. □

We briefly discuss a simple application of the preferential attachment model in the following text.

Bootstrap Percolation

This is a simplified mathematical model of the spread of a disease through a graph/network $G = (V, E)$. Initially a set A_0 of vertices are considered to be infected. This is considered to be round 0. Then in round $t > 0$, any vertex that has at least r neighbors in A_{t-1} will become infected. No-one recovers in this model. The main question is as to how many vertices eventually end up getting infected. There is a large literature on this subject with a variety of graphs G and ways of defining A_0.

Here we will assume that each vertex s is placed in A_0 with probability p, independent of other vertices. The proof of the following theorem relies on the fact that with high probability all of the early vertices of G_t become infected during the first round. Subsequently, the connectivity of the random graph is enough to spread the infection to the remaining vertices. The following is a simplified version of Theorem 1 of Abdullah and Fountoulakis [1].

Theorem 12.8 *If* $r \leq m$ *and* $\omega = \log^2 t$ *and* $p \geq \omega t^{-1/2}$ *then w.h.p. all vertices in* G_t *get infected.*

Proof Given Theorem 12.7, we can assume that $d_s(t) \geq mt^{1/2}/\omega^{1/2}$ for $1 \leq s \leq m$. In which case, the probability that vertex $s \leq m$ is not infected in round 1 is at most

$$\sum_{i=1}^{m-1} \binom{mt^{1/2}/\omega^{1/2}}{i} p^i (1-p)^{mt^{1/2}/\omega^{1/2}-i} \leq \sum_{i=1}^{m-1} \omega^{i/2} e^{-(1-o(1))m\omega^{1/2}} = o(1). \quad (12.19)$$

So, w.h.p. $1, 2, \ldots, m$ are infected in round 1. After this we use induction and the fact that every vertex $i > s$ has m neighbors $j < i$. □

Exercises

12.1.1 Verify equation (12.8).

12.1.2 Verify equation (12.15).

12.1.3 Verify equation (12.19).

12.2 Spatial Preferential Attachment

The Spatial Preferential Attachment (SPA) model was introduced by Aiello, Bonato, Cooper, Janssen, and Prałat in [2]. This model combines preferential attachment with geometry by introducing "spheres of influence" of vertices, whose volumes depend on their in-degrees.

We first fix parameters of the model. Let $m \in \mathbb{N}$ be the dimension of space $\mathbb{R}^m$, $p \in [0, 1]$ be the link (arc) probability, and fix two additional parameters A_1, A_2, where $A_1 < 1/p$ while $A_2 > 0$. Let S be the unit hypercube in $\mathbb{R}^m$, with the torus metric $d(\cdot, \cdot)$ derived from the L^∞ metric. In particular, for any two points x and y in S,

$$d(x, y) = \min \{\|x - y + u\|_\infty : u \in \{-1, 0, 1\}^m\}. \quad (12.20)$$

For each positive real number $\alpha < 1$, and $u \in S$, define the ball around u with volume α as

$$B_\alpha(u) = \{x \in S : d(u, x) \leq r_\alpha\},$$

where $r_\alpha = \alpha^{1/m}/2$, so that r_α is chosen such that B_α has volume α.

The SPA model generates a stochastic sequence of directed graphs $\{G_t\}$, where $G_t = (V_t, E_t)$ and $V_t \subset S$, i.e., all vertices are placed in the m-dimensional hypercube $S = [0, 1]^m$.

Let $\deg^-(v; t)$ be the in-degree of the vertex v in G_t, and $\deg^+(v; t)$ its out-degree.

Then, the *sphere of influence* $S(v;t)$ of the vertex v at time $t \geq 1$ is the ball centered at v with the following volume:

$$|S(v,t)| = \min\left\{\frac{A_1 \deg^-(v;t) + A_2}{t}, 1\right\}. \tag{12.21}$$

In order to construct a sequence of graphs we start at $t = 0$ with G_0 being the null graph. At each time step t we construct G_t from G_{t-1} by, first, choosing a new vertex v_t uniformly at random (**uar**) from the cube S and adding it to V_{t-1} to create V_t. Then, independently, for each vertex $u \in V_{t-1}$ such that $v_t \in S(u, t-1)$, a directed link (v_t, u) is created with probability p. Thus, the probability that a link (v_t, u) is added in time step t equals $p|S(u, t-1)|$.

Power Law and Vertex In-degrees

Theorem 12.9 *Let $N_{i,n}$ be the number of vertices of in-degree i in the SPA graph G_t at time $t = n$, where $n \geq 0$ is an integer. Fix $p \in (0, 1]$. Then for any $i \geq 0$,*

$$\mathbb{E}(N_{i,n}) = (1 + o(1))c_i n, \tag{12.22}$$

where

$$c_0 = \frac{1}{1 + pA_2}, \tag{12.23}$$

and for $1 \leq i \leq n$,

$$c_i = \frac{p^i}{1 + pA_2 + ipA_1} \prod_{j=0}^{i-1} \frac{jA_1 + A_2}{1 + pA_2 + jpA_1}. \tag{12.24}$$

In [2] a stronger result is proved which indicates that the fraction $N_{i,n}/n$ follows a power law. It is shown that for $i = 0, 1, \ldots, i_f$, where $i_f = (n/\log^8 n)^{pA_1/(4pA_1+2)}$, w.h.p.

$$N_{i,n} = (1 + o(1))c_i n.$$

Since, for some constant c,

$$c_i = (1 + o(1))ci^{-(1+1/pA_1)},$$

it shows that for large i the expected proportion $N_{i,n}/n$ follows a power law with exponent $1 + \frac{1}{pA_1}$, and concentration for all values of i up to i_f.

To prove Theorem 12.9 we need the following result of Chung and Lu (see [32], Lemma 3.1) on real sequences.

Lemma 12.10 *Let $\{\alpha_t\}$, $\{\beta_t\}$ and $\{\gamma_t\}$ be real sequences satisfying the relation*

$$\alpha_{t+1} = \left(1 - \frac{\beta_t}{t}\right)\alpha_t + \gamma_t.$$

Furthermore, suppose $\lim_{t\to\infty}\beta_t = \beta > 0$ *and* $\lim_{t\to\infty}\gamma_t = \gamma$. *Then* $\lim_{t\to\infty}\frac{\alpha_t}{t}$ *exists and*

$$\lim_{t\to\infty}\frac{\alpha_t}{t} = \frac{\gamma}{1+\beta}.$$

□

Proof of Theorem 12.9

The equations relating the random variables $N_{i,t}$ are described as follows. Since G_1 consists of one isolated node, $N_{0,1} = 1$, and $N_{i,1} = 0$ for $i > 0$. For all $t > 0$, we derive that

$$\mathbb{E}(N_{0,t+1} - N_{0,t}|G_t) = 1 - pN_{0,t}\frac{A_2}{t}, \tag{12.25}$$

while

$$\mathbb{E}(N_{i,t+1} - N_{i,t}|G_t) = pN_{i-1,t}\frac{A_1 i + A_2}{t} - pN_{i,t}\frac{A_1(i-1)+A_2}{t}. \tag{12.26}$$

Now applying Lemma 12.10 to (12.25) with

$$\alpha_t = \mathbb{E}(N_{0,t})\ \ \beta_t = pA_2 \text{ and } \gamma_t = 1,$$

we get that

$$\mathbb{E}(N_{0,t}) = c_0 + o(t),$$

where c_0 as in (12.23).

For $i > 0$, Lemma 12.10 can be inductively applied with

$$\alpha_t = \mathbb{E}(N_{i,t}),\ \ \beta_t = p(A_1 i + A_2) \text{ and } \gamma_t = \mathbb{E}(N_{i-1,t})\frac{A_1(i-1)+A_2}{t}$$

to show that

$$\mathbb{E}(N_{i,t}) = c_i + o(t),$$

where

$$c_i = pc_{i-1}\frac{A_1(i-1)+A_2}{1+p(A_1 i + A_2)}.$$

One can easily verify that the expressions for c_0, and c_i, $i \geq 1$, given in (12.23) and (12.24), satisfy the respective recurrence relations derived above. □

Knowing the expected in-degree of a node, given its age, can be used to analyze geometric properties of the SPA graph G_t. Let us note also that the result below for $i \gg 1$ was proved in [61] and extended to all $i \geq 1$ in [37]. As before, let v_i be the node added at time i.

Theorem 12.11 *Suppose that* $i = i(t) \gg 1$ *as* $t \to \infty$. *Then,*

$$\mathbb{E}(\deg^-(v_i,t)) = (1+o(1))\frac{A_2}{A_1}\left(\frac{t}{i}\right)^{pA_1} - \frac{A_2}{A_1},$$

$$\mathbb{E}(|S(v_i,t)|) = (1+o(1))A_2 t^{pA_1-1} i^{-pA_1}. \tag{12.27}$$

Moreover, for all $i \geq 1$,

$$\mathbb{E}(\deg^-(v_i,t)) \leq \frac{eA_2}{A_1}\left(\frac{t}{i}\right)^{pA_1} - \frac{A_2}{A_1},$$
$$\mathbb{E}(|S(v_i,t)|) \leq (1+o(1))eA_2 t^{pA_1-1} i^{-pA_1}. \tag{12.28}$$

Proof In order to simplify calculations, we make the following substitution:

$$X(v_i,t) = \deg^-(v_i,t) + \frac{A_2}{A_1}. \tag{12.29}$$

It follows from the definition of the process that

$$X(v_i,t+1) = \begin{cases} X(v_i,t)+1, & \text{with probability } \frac{pA_1X(v_i,t)}{t}, \\ X(v_i,t), & \text{otherwise.} \end{cases}$$

We then have

$$\begin{aligned}
&\mathbb{E}(X(v_i,t+1) \mid X(v_i,t)) \\
&= (X(v_i,t)+1)\frac{pA_1X(v_i,t)}{t} + X(v_i,t)\left(1 - \frac{pA_1X(v_i,t)}{t}\right) \\
&= X(v_i,t)\left(1+\frac{pA_1}{t}\right).
\end{aligned}$$

Taking expectations over $X(v_i,t)$, we get

$$\mathbb{E}(X(v_i,t+1)) = \mathbb{E}(X(v_i,t))\left(1+\frac{pA_1}{t}\right).$$

Since all nodes start with in-degree zero, $X(v_i,i) = \frac{A_2}{A_1}$. Note that, for $0 < x < 1$, $\log(1+x) = x - O(x^2)$. If $i \gg 1$, one can use this to get

$$\mathbb{E}(X(v_i,t)) = \frac{A_2}{A_1}\prod_{j=i}^{t-1}\left(1+\frac{pA_1}{j}\right) = (1+o(1))\frac{A_2}{A_1}\exp\left(\sum_{j=i}^{t-1}\frac{pA_1}{j}\right),$$

and in all cases $i \geq 1$,

$$\mathbb{E}(X(v_i,t)) \leq \frac{A_2}{A_1}\exp\left(\sum_{j=i}^{t-1}\frac{pA_1}{j}\right).$$

Therefore, when $i \gg 1$,

$$\mathbb{E}(X(v_i,t)) = (1+o(1))\frac{A_2}{A_1}\exp\left(pA_1\log\left(\frac{t}{i}\right)\right) = (1+o(1))\frac{A_2}{A_1}\left(\frac{t}{i}\right)^{pA_1},$$

and (12.27) follows from (12.29) and (12.21). Moreover, for any $i \geq 1$,

$$\mathbb{E}(X(v_i,t)) \leq \frac{A_2}{A_1}\exp\left(pA_1\left(\log\left(\frac{t}{i}\right)+1/i\right)\right) \leq \frac{eA_2}{A_1}\left(\frac{t}{i}\right)^{pA_1},$$

and (12.28) follows from (12.29) and (12.21) as before, which completes the proof. □

Directed Diameter

Consider the graph G_t produced by the SPA model. For a given pair of vertices $v_i, v_j \in V_t$ $(1 \leq i < j \leq t)$, let $l(v_i, v_j)$ denote the length of the shortest directed path from v_j to v_i if such a path exists, and let $l(v_i, v_j) = 0$ otherwise. The directed diameter of a graph G_t is defined as

$$\mathrm{diam}(G_t) = \max_{1 \leq i < j \leq t} l(v_i, v_j).$$

We next prove the following upper bound on $\mathrm{diam}(G_t)$ (see [37]):

Theorem 12.12 *Consider the SPA model. There exists an absolute constant $c_1 > 0$ such that w.h.p.*

$$\mathrm{diam}(G_t) \leq c_1 \log t.$$

Proof Let $C = 18 \max(A_2, 1)$. We prove that with probability $1 - o(t^{-2})$ we have that for any $1 \leq i < j \leq t$, G_t does not contain a directed (v_i, v_j)-path of length exceeding $k^* = C \log t$. As there are at most t^2 pairs v_i, v_j, Theorem 12.12 will follow.

In order to simplify the notation, we use v to denote the vertex added at step $v \leq t$. Let $vPu = (v, t_{k-1}, t_{k-2}, \ldots, t_1, u)$ be a directed (v, u)-path of length k where $t_0 = u$, $t_k = v$.

$$\mathbb{P}(vPu \text{ exists}) = \prod_{i=1}^{k} p \left(\frac{A_1 \deg^-(t_{i-1}, t_i) + A_2}{t_i} \right).$$

Let $N(v, u, k)$ be the number of directed (v, u)-paths of length k, then

$$\mathbb{E}(N(v, u, k)) = \sum_{u < t_1 < \cdots < t_{k-1} < v} p^k \, \mathbb{E}\left(\prod_{i=1}^{k} \left(\frac{A_1 \deg^-(t_{i-1}, t_i) + A_2}{t_i} \right) \right).$$

We first consider the case where u tends to infinity together with t. It follows from Theorem 12.11 that

$$\mathbb{E}(\deg^-(t_{i-1}, t_i)) = (1 + o(1)) \frac{A_2}{A_1} \left(\frac{t_i}{t_{i-1}} \right)^{pA_1} - \frac{A_2}{A_1}.$$

Thus

$$\begin{aligned}
\mathbb{E}(N(v, u, k)) &= \sum_{u < t_1 < \cdots < t_{k-1} < v} p^k \prod_{i=1}^{k} \frac{1}{t_i} (A_1 \, \mathbb{E}(\deg^-(t_{i-1}, t_i)) + A_2) \\
&= \sum_{u < t_1 < \cdots < t_{k-1} < v} (1 + o(1))^k (A_2 p)^k \prod_{i=1}^{k} \frac{1}{t_i} \left(\frac{t_i}{t_{i-1}} \right)^{pA_1} \\
&= (1 + o(1))^k (A_2 p)^k \left(\frac{v}{u} \right)^{pA_1} \frac{1}{v} \sum_{u < t_1 < \cdots < t_{k-1} < v} \prod_{i=1}^{k-1} \frac{1}{t_i}.
\end{aligned}$$

However,

$$\sum_{u<t_1<\cdots<t_{k-1}<v}\prod_{i=1}^{k-1}\frac{1}{t_i} \le \frac{1}{(k-1)!}\left(\sum_{u<s<v}\frac{1}{s}\right)^{k-1}$$
$$\le \frac{1}{(k-1)!}(\log v/u + 1/u)^{k-1}$$
$$\le \left(\frac{e(\log v/u + 1/u)}{k-1}\right)^{k-1}.$$

Let $k^* = C\log t$, where $C = 18\max(1, A_2)$. Assuming that t is sufficiently large, and recalling that $pA_1 < 1$, we have

$$\sum_{k>k^*}\mathbb{E}(N(v,u,k)) \le 2A_2\sum_{k>k^*}\left(\frac{(1+o(1))A_2pe(\log v/u + 1/u)}{k-1}\right)^{k-1}$$
$$\le 2A_2\left(\frac{(1+o(1))A_2e(\log v/u + 1/u)}{C\log t}\right)^{k^*}\frac{1}{1-3A_2/C} \tag{12.30}$$
$$= O(6^{-18\log t}) \tag{12.31}$$
$$= o(t^{-4}).$$

The result follows for u tending to infinity. In the case where u is a constant, it follows from Theorem 12.11 that a multiplicative correction of e can be used in $\mathbb{E}(\deg^-(t_{i-1}, t_i))$, leading to replacing e by e^2 in (12.30) and then 6 in (12.31) by 2, giving a bound of $O(2^{-18\log t}) = o(t^{-4})$ as before. This finishes the proof of the theorem. □

Exercises

12.2.1 Prove Lemma 12.10.

12.2.2 Verify equation (12.30).

12.2.3 Verify equation (12.31).

Problems for Chapter 12

12.1 Let G_t be the preferential attachment graph defined in Section 12.1 with $m = 1$, i.e., G_t is a plane-oriented recursive tree on t vertices. Let L_t denote the number of leaves (vertices of degree one) of G_t. Show that

$$\mathbb{E}\,L_t = \frac{2t-1}{3} \quad\text{and}\quad \operatorname{Var} L_t = \frac{2t(t-2)}{9(2t-3)}.$$

12.2 Prove that L_t/t (defined above) converges in probability, to $2/3$.

12.3 Show that w.h.p. the Preferential Attachment Graph of Section 12.1 has diameter $O(\log n)$. (Hint: Using the idea that vertex t chooses a random edge of the current graph, observe that half of these edges appeared at time $t/2$ or less).

12.4 For the next few questions, we modify the Preferential Attachment Graph of Section 12.1 in the following way: First, let $m = 1$ and preferentially generate a sequence of graphs $\Gamma_1, \Gamma_2, \ldots, \Gamma_{mn}$. Then if the edges of Γ_{mn} are $(u_i, v_i), i = 1, 2, \ldots, mn$, let the edges of $\mathbb{G}_n$ be $(u_{\lceil i/m \rceil}, v_{\lceil i/m \rceil}), i = 1, 2, \ldots, mn$. Show that (12.1) continues to hold.

12.5 Show that $\mathbb{G}_n$ of the previous question can also be generated in the following way:

(a) Let π be a random permutation of $[2mn]$. Let $X = \{(a_i, b_i), i = 1, 2, \ldots, mn\}$ where $a_i = \min\{\pi(2i-1), \pi(2i)\}$ and $b_i = \max\{\pi(2i-1), \pi(2i)\}$.

(b) Let the edges of $\mathbb{G}_n$ be $(a_{\lceil i/m \rceil}, b_{\lceil i/m \rceil}), i = 1, 2, \ldots, mn$.

This model was introduced in [28].

12.6 Show that the edges of the graph in the previous question can be generated as follows:

(a) Let $\zeta_1, \zeta_2, \ldots, \zeta_{2mn}$ be independent uniform $[0, 1]$ random variables. Let $\{x_i < y_i\} = \{\zeta_{2i-1}, \zeta_{2i}\}$ for $i = 1, 2, \ldots, mn$. Sort the y_i in increasing order $R_1 < R_2 < \cdots < R_{mn}$ and let $R_0 = 0$. Then let

$$W_j = R_{mj} \text{ and } I_j = (W_{j-1}, W_j] \text{ for } j = 1, 2, \ldots, n.$$

This model was introduced in [27].

(b) The edges of $\mathbb{G}_n$ are $(u_i, v_i), i = 1, 2, \ldots, mn$ where $x_i \in I_{u_i}, y_i \in I_{v_i}$.

12.7 Prove that $(R_1, R_2, \ldots, R_{mn})$ can be generated as

$$R_i = \left(\frac{\Upsilon_i}{\Upsilon_{mn+1}} \right)^{1/2}$$

where $\Upsilon_N = \xi_1 + \xi_2 + \cdots + \xi_N$ for $N \geq 1$ and $\xi_1, \xi_2, \ldots, \xi_{mn+1}$ are independent exponential copies of $EXP(1)$.

12.8 Let L be a large constant and let $\omega = \omega(n) \to \infty$ arbitrarily slowly. Then let $\mathcal{E}$ be the event that

$$\Upsilon_k \sim k \text{ for } \frac{k}{m} \in [\omega, n] \text{ or } k = mn + 1.$$

Show that

(a) $\mathbb{P}(\neg\mathcal{E}) = o(1)$.

(b) Let $\eta_i = \xi_{(i-1)m+1} + \xi_{(i-1)m+2} + \cdots + \xi_{im}$. If $\mathcal{E}$ occurs then

(1) $W_i \sim \left(\frac{i}{n} \right)^{1/2}$ for $\omega \leq i \leq n$, and

(2) $w_i = W_i - W_{i-1} \sim \frac{\eta_i}{2m(in)^{1/2}}$ for $\omega \leq i \leq n$.

(c) $\eta_i \leq \log n$ for $i \in [n]$ w.h.p.

(d) $\eta_i \leq \log\log n$ for $i \in [(\log n)^{10}]$ w.h.p.

(e) If $\omega \leq i < j \leq n$ then $\mathbb{P}(\text{edge } ij \text{ exists}) \sim \frac{\eta_i}{2(ij)^{1/2}}$.

(f) $\eta_i \geq \frac{1}{\log\log n}$ and $i \leq \frac{n}{\omega(\log n)^3}$ implies the degree $d_n(i) \sim \eta_i \left(\frac{n}{i}\right)^{1/2}$.

13 Intersection Graphs

Let G be a (finite, simple) graph. We say that G is an *intersection graph* if we can assign to each vertex $v \in V(G)$ a set S_v so that $\{v, w\} \in E(G)$ exactly when $S_v \cap S_w \neq \emptyset$. In this case, we say G is the intersection graph of the family of sets $\mathcal{S} = \{S_v : v \in V(G)\}$.

Although all graphs are intersection graphs (see Marczewski [84]) some classes of intersection graphs are of special interest. Depending on the choice of family $\mathcal{S}$, often reflecting some geometric configuration, one can consider, for example, *interval graphs* defined as the intersection graphs of intervals on the real line, *unit disc graphs* defined as the intersection graphs of unit discs on the plane, etc. In this chapter, we will discuss properties of *random intersection graphs*, where the family $\mathcal{S}$ is generated in a random manner.

It is worth mentioning that specific variants of random intersection graphs have a diverse range of applications. For example, binomial intersection random graphs, the topic of the first section, can serve as a model of networks with communities, i.e., local structures (cliques, bipartite cliques, etc.) with higher density than the network average. They can also be applied in classification to find clusters and to test their randomness in sets of nonmetric data, as well as in studying techniques of handling incomplete (missing) data, just to name a few.

Random geometric graphs, studied in Section 13.2, find a huge number of applications in modeling telecommunication (cell phone), information, security, neuronal, and many other real-world networks.

13.1 Binomial Random Intersection Graphs

Binomial random intersection graphs were introduced by Karoński, Scheinerman, and Singer-Cohen [68] as a generalization of the classical model of the binomial random graph $\mathbb{G}_{n,p}$.

Let n, m be positive integers and let $0 \leq p \leq 1$. Let $V = \{1, 2, \ldots, n\}$ be the set of vertices and for every $1 \leq k \leq n$, let S_k be a random subset of the set $M = \{1, 2, \ldots, m\}$ formed by selecting each element of M independently with probability p. We define a *binomial random intersection graph* $G(n, m, p)$ as the intersection graph of sets $S_k, \ k = 1, 2, \ldots, n$. Here $S_1, S_2, \ldots, S_n$ are generated independently. Hence, two vertices i and j are adjacent in $G(n, m, p)$ if and only if $S_i \cap S_j \neq \emptyset$.

There are other ways to generate binomial random intersection graphs. For example, we may start with a classical bipartite random graph $\mathbb{G}_{n,m,p}$, with vertex set bipartition

$$(V, M), V = \{1, 2, \ldots, n\}, M = \{1, 2, \ldots, m\},$$

where each edge between V and M is drawn independently with probability p. Next, one can generate a graph $G(n, m, p)$ with vertex set V and vertices i and j of $G(n, m, p)$ connected if and only if they share a common neighbor (in M) in the random graph $\mathbb{G}_{n,m,p}$. Here the graph $\mathbb{G}_{n,m,p}$ is treated as a *generator* of $G(n, m, p)$.

One observes that the probability that there is an edge $\{i, j\}$ in $G(n, m, p)$ equals $1-(1-p^2)^m$, since the probability that sets S_i and S_j are disjoint is $(1-p^2)^m$; however, in contrast with $\mathbb{G}_{n,p}$, the edges do not occur independently of each other.

Another simple observation leads to some natural restrictions on the choice of probability p. Note that the expected number of edges of $G(n, m, p)$ is

$$\binom{n}{2}(1 - (1 - p^2)^m) \sim n^2 m p^2/2,$$

provided $mp^2 \to 0$ as $n \to \infty$. Therefore, if we take $p = o((n\sqrt{m})^{-1})$, then the expected number of edges of $G(n, m, p)$ tends to 0 as $n \to \infty$ and therefore w.h.p. $G(n, m, p)$ is empty.

On the other hand, the expected number of non-edges in $G(n, m, p)$ is

$$\binom{n}{2}(1 - p^2)^m \leq n^2 e^{-mp^2}.$$

Thus, if we take $p = (2 \log n + \omega(n))/m)^{1/2}$, where $\omega(n) \to \infty$ as $n \to \infty$, then the random graph $G(n, m, p)$ is complete w.h.p. One can also easily show that when $\omega(n) \to -\infty$, then $G(n, m, p)$ is w.h.p. not complete. So, when studying the evolution of $G(n, m, p)$, we may restrict ourselves to values of p in the range between $\omega(n)/(n\sqrt{m})$ and $((2 \log n - \omega(n))/m)^{1/2}$, where $\omega(n) \to \infty$.

Equivalence

One of the first interesting problems to be considered is the question as to when the random graphs $G(n, m, p)$ and $\mathbb{G}_{n,p}$ have asymptotically the same properties. Intuitively, it should be the case when the edges of $G(n, m, p)$ occur "almost independently," i.e., when there are no vertices of degree greater than two in M in the generator $\mathbb{G}_{n,m,p}$ of $G(n, m, p)$. Then each of its edges is induced by a vertex of degree two in M, "almost" independently of other edges. One can show that this happens w.h.p. when $p = o\left(1/(nm^{1/3})\right)$, which in turn implies that both random graphs are asymptotically equivalent for all graph properties $\mathcal{P}$. Recall that a graph property $\mathcal{P}$ is defined as a subset of the family of all labeled graphs on vertex set $[n]$, i.e., $\mathcal{P} \subseteq 2^{\binom{n}{2}}$. The following equivalence result is due to Rybarczyk [104] and Fill, Scheinerman, and Singer-Cohen [48].

Theorem 13.1 *Let* $0 \leq a \leq 1$, $\mathcal{P}$ *be any graph property,* $p = o\left(1/(nm^{1/3})\right)$ *and*

$$\hat{p} = 1 - \exp\left(-mp^2(1-p)^{n-2}\right). \tag{13.1}$$

Then

$$\mathbb{P}(\mathbb{G}_{n,\hat{p}} \in \mathcal{P}) \to a$$

if and only if

$$\mathbb{P}(G(n, m, p) \in \mathcal{P}) \to a$$

as $n \to \infty$.

Proof Let X and Y be random variables taking values in a common finite (or countable) set S. Consider the probability measures $\mathcal{L}(X)$ and $\mathcal{L}(Y)$ on S whose values at $A \subseteq S$ are $\mathbb{P}(X \in A)$ and $\mathbb{P}(Y \in A)$. Define the total variation distance between $\mathcal{L}(X)$ and $\mathcal{L}(Y)$ as

$$d_{TV}(\mathcal{L}(X), \mathcal{L}(Y)) := \sup_{A \subseteq S} |\mathbb{P}(X \in A) - \mathbb{P}(Y \in A)|,$$

which is equivalent to

$$d_{TV}(\mathcal{L}(X), \mathcal{L}(Y)) = \frac{1}{2} \sum_{s \in S} |\mathbb{P}(X = s) - \mathbb{P}(Y = s)|.$$

Note (see Fact 4 of [48]) that if there exists a probability space on which random variables X' and Y' are both defined, with $\mathcal{L}(X) = \mathcal{L}(X')$ and $\mathcal{L}(Y) = \mathcal{L}(Y')$, then

$$d_{TV}(\mathcal{L}(X), \mathcal{L}(Y)) \leq \mathbb{P}(X' \neq Y'). \tag{13.2}$$

Furthermore (see Fact 3 of [48]), if there exist random variables Z and Z' such that $\mathcal{L}(X|Z = z) = \mathcal{L}(Y|Z' = z)$ for all z, then

$$d_{TV}(\mathcal{L}(X), \mathcal{L}(Y)) \leq 2d_{TV}(\mathcal{L}(Z), \mathcal{L}(Z')). \tag{13.3}$$

We will need one more observation. Suppose that a random variable X has the distribution $\text{Bin}(n, p)$, while a random variable Y has the Poisson distribution, and $\mathbb{E}\, X = \mathbb{E}\, Y$. Then

$$d_{TV}(X, Y) = O(p). \tag{13.4}$$

We leave the proofs of (13.2), (13.3), and (13.4) as problems listed at the end of this chapter.

To prove Theorem 13.1, we also need some auxiliary results on a special coupon collector scheme.

Let Z be a non-negative integer-valued random variable, r a non-negative integer, and γ a real such that $r\gamma \leq 1$. Assume we have r coupons $Q_1, Q_2, \dots, Q_r$ and one blank coupon B. We make Z independent draws (with replacement) such that in each draw,

$$\mathbb{P}(Q_i \text{ is chosen}) = \gamma, \quad \text{for } i = 1, 2, \dots, r,$$

and

$$\mathbb{P}(B \text{ is chosen}) = 1 - r\gamma.$$

Let $N_i(Z)$, $i = 1, 2, \dots, r$ be a random variable counting the number of times that coupon Q_i was chosen. Furthermore, let

$$X_i(Z) = \begin{cases} 1 & \text{if } N_i(Z) \geq 1, \\ 0 & \text{otherwise.} \end{cases}$$

The number of different coupons selected is given by

$$X(Z) = \sum_{i=1}^{r} X_i(Z). \tag{13.5}$$

With the above-mentioned definitions we observe that the following holds.

Lemma 13.2 *If a random variable Z has the Poisson distribution with expectation λ, then $N_i(Z)$, $i = 1, 2, \dots, r$, are independent and identically Poisson distributed random variables, with expectation $\lambda\gamma$. Moreover, the random variable $X(Z)$ has the distribution* $\mathrm{Bin}(r, 1 - e^{-\lambda\gamma})$.

Let us consider the following special case of the scheme defined earlier, assuming that $r = \binom{n}{2}$ and $\gamma = 1/\binom{n}{2}$. Here each coupon represents a distinct edge of K_n.

Lemma 13.3 *Suppose $p = o(1/n)$ and let random variable Z be $\mathrm{Bin}\left(m, \binom{n}{2}p^2(1-p)^{n-2}\right)$ distributed, while random variable Y is $\mathrm{Bin}\left(\binom{n}{2}, 1 - e^{-mp^2(1-p)^{n-2}}\right)$ distributed, then*

$$d_{TV}(\mathcal{L}(Y), \mathcal{L}(X(Z))) = o(1).$$

Proof Let Z' be a Poisson random variable with the same expectation as Z, i.e.,

$$\mathbb{E}\, Z' = m\binom{n}{2}p^2(1-p)^{n-2}.$$

By Lemma 13.2, $X(Z')$ has the binomial distribution

$$\mathrm{Bin}\left(\binom{n}{2}, 1 - e^{-mp^2(1-p)^{n-2}}\right),$$

and so, by (13.3) and (13.4), we have

$$\begin{aligned} d_{TV}(\mathcal{L}(Y), \mathcal{L}(X(Z))) &= d_{TV}(\mathcal{L}(X(Z')), \mathcal{L}(X(Z))) \\ &\le 2d_{TV}(\mathcal{L}(Z'), \mathcal{L}(Z)) \le O\left(\binom{n}{2} p^2 (1-p)^{n-2}\right) \\ &= O\left(n^2 p^2\right) = o(1). \end{aligned}$$

□

Now define a random intersection graph $G_2(n, m, p)$ as follows. Its vertex set is $V = \{1, 2, \ldots, n\}$, while $e = \{i, j\}$ is an edge in $G_2(n, m, p)$ if and only if in a (generator) bipartite random graph $\mathbb{G}_{n,m,p}$ there is a vertex $w \in M$ of degree two such that both i and j are connected by an edge with w.

To complete the proof of our theorem, note that

$$\begin{aligned} &d_{TV}(\mathcal{L}(G(n, m, p)), \mathcal{L}(\mathbb{G}_{n,\hat{p}})) \\ &\quad \le d_{TV}(\mathcal{L}(G(n, m, p)), \mathcal{L}(G_2(n, m, p))) + d_{TV}(\mathcal{L}(G_2(n, m, p)), \mathcal{L}(\mathbb{G}_{n,\hat{p}})), \end{aligned}$$

where $\hat{p}$ is defined in (13.1). Now, by (13.2),

$$\begin{aligned} &d_{TV}(\mathcal{L}(G(n, m, p)), \mathcal{L}(G_2(n, m, p))) \\ &\qquad \le \mathbb{P}(\mathcal{L}(G(n, m, p)) \neq \mathcal{L}(G_2(n, m, p))) \\ &\qquad \le \mathbb{P}(\exists w \in M \text{ of } \mathbb{G}_{n,m,p} \text{ s.t. } deg(w) > 2) \le m\binom{n}{3} p^3 = o(1) \end{aligned}$$

for $p = o(1/(nm^{1/3}))$.

Hence, it remains to show that

$$d_{TV}(\mathcal{L}(G_2(n, m, p)), \mathcal{L}(\mathbb{G}_{n,\hat{p}})) = o(1). \tag{13.6}$$

Let Z be distributed as $\text{Bin}\left(m, \binom{n}{2} p^2 (1-p)^{n-2}\right)$, $X(Z)$ is defined as in (13.5), and let Y be distributed as $\text{Bin}\left(\binom{n}{2}, 1 - e^{-mp^2(1-p)^{n-2}}\right)$. Then the number of edges $|E(G_2(n, m, p))| = X(Z)$ and $|E(\mathbb{G}_{n,\hat{p}}))| = Y$. Moreover, for any two graphs G and G' with the same number of edges

$$\mathbb{P}(G_2(n, m, p) = G) = \mathbb{P}(G_2(n, m, p) = G')$$

and

$$\mathbb{P}(\mathbb{G}_{n,\hat{p}} = G) = \mathbb{P}(\mathbb{G}_{n,\hat{p}} = G').$$

Equation (13.6) now follows from Lemma 13.3. The theorem follows immediately. □

For monotone properties (see Section 3.1), the relationship between the classical binomial random graph and the respective intersection graph is more precise and was established by Rybarczyk [104].

Theorem 13.4 *Let $0 \le a \le 1$, $m = n^\alpha, \alpha \ge 3$. Let $\mathcal{P}$ be any monotone graph property. For $\alpha > 3$, assume*

$$\Omega\left(1/(nm^{1/3})\right) = p = O(\sqrt{\log n/m}),$$

while for $\alpha = 3$, *assume* $\left(1/(nm^{1/3})\right) = o(p)$. *Let*

$$\hat{p} = 1 - \exp\left(-mp^2(1-p)^{n-2}\right).$$

If for all $\varepsilon = \varepsilon(n) \to 0$,

$$\mathbb{P}(\mathbb{G}_{n,(1+\varepsilon)\hat{p}} \in \mathcal{P}) \to a,$$

then

$$\mathbb{P}(G(n,m,p) \in \mathcal{P}) \to a$$

as $n \to \infty$.

Small Subgraphs

Let H be any fixed graph. A *clique cover* C is a collection of subsets of vertex set $V(H)$ such that each induces a complete subgraph (*clique*) of H, and for every edge $\{u,v\} \in E(H)$, there exists $C \in C$ such that $u, v \in C$. Hence, the cliques induced by sets from C exactly cover the edges of H. A clique cover is allowed to have more than one copy of a given set. We say that C is *reducible* if for some $C \in C$, the edges of H induced by C are contained in the union of the edges induced by $C \backslash C$, otherwise C is *irreducible*. Note that if $C \in C$ and C is irreducible, then $|C| \geq 2$.

In this section, $|C|$ stands for the number of cliques in C, while $\sum C$ denotes the sum of clique sizes in C, and we put $\sum C = 0$ if $C = \emptyset$.

Let $C = \{C_1, C_2, \ldots, C_k\}$ be a clique cover of H. For $S \subseteq V(H)$, define the following two *restricted clique covers*

$$C_t[S] := \{C_i \cap S : |C_i \cap S| \geq t,\ \ i = 1, 2, \ldots, k\},$$

where $t = 1, 2$. For a given S and $t = 1, 2$, let

$$\tau_t = \tau_t(H, C, S) = \left(n^{|S|/\sum C_t[S]} m^{|C_t[S]|/\sum C_t[S]}\right)^{-1}.$$

Finally, let

$$\tau(H) = \min_{C} \max_{S \subseteq V(H)} \{\tau_1, \tau_2\},$$

where the minimum is taken over all clique covers C of H. We can in this calculation restrict our attention to irreducible covers.

Karoński, Scheinerman, and Singer-Cohen [68] proved the following theorem.

Theorem 13.5 *Let H be a fixed graph and $mp^2 \to 0$. Then*

$$\lim_{n\to\infty} \mathbb{P}(H \subseteq G(n,m,p)) = \begin{cases} 0 & \text{if } p/\tau(H) \to 0, \\ 1 & \text{if } p/\tau(H) \to \infty. \end{cases}$$

As an illustration, we will use this theorem to show the threshold for complete graphs in $G(n,m,p)$, when $m = n^{\alpha}$, for different ranges of $\alpha > 0$.

Corollary 13.6 *For a complete graph K_h with $h \geq 3$ vertices and $m = n^\alpha$, we have*

$$\tau(K_h) = \begin{cases} n^{-1}m^{-1/h} & \text{for } \alpha \leq 2h/(h-1), \\ n^{-1/(h-1)}m^{-1/2} & \text{for } \alpha \geq 2h/(h-1). \end{cases}$$

Proof There are many possibilities for clique covers to generate a copy of a complete graph K_h in $G(n, m, p)$. However, in the case of K_h, only two play a dominating role. Indeed, we will show that for $\alpha \leq \alpha_0$, $\alpha_0 = 2h/(h-1)$, the clique cover $C = \{V(K_h)\}$ composed of one set containing all h vertices of K_h only matters, while for $\alpha \geq \alpha_0$ the clique cover $C = \binom{K_h}{2}$, consisting of $\binom{h}{2}$ pairs of endpoints of the edges of K_h, takes the leading role.

Let $V = V(K_h)$ and denote those two clique covers by $\{V\}$ and $\{E\}$, respectively. Observe that for the cover $\{V\}$ the following equality holds:

$$\max_{S \subseteq V}\{\tau_1(K_h, \{V\}, S), \tau_2(K_h, \{V\}, S)\} = \tau_1(K_h, \{V\}, V). \tag{13.7}$$

To see this, check first that for $|S| = h$,

$$\tau_1(K_h, \{V\}, V) = \tau_2(K_h, \{V\}, V) = n^{-1}m^{-1/h}.$$

For S of size $|S| = s$, $2 \leq s \leq h-1$, restricting the clique cover $\{V\}$ to S gives a single s-clique, so for $t = 1, 2$,

$$\tau_t(K_h, \{V\}, S) = n^{-1}m^{-1/s} < n^{-1}m^{-1/h}.$$

Finally, when $|S| = 1$, then $\tau_1 = (nm)^{-1}$, while $\tau_2 = 0$, both smaller than $n^{-1}m^{-1/h}$, and so equation (13.7) follows.

For the edge-clique cover $\{E\}$, we have a similar expression, namely.

$$\max_{S \subseteq V}\{\tau_1(K_h, \{E\}, S), \tau_2(K_h, \{E\}, S)\} = \tau_1(K_h, \{E\}, V). \tag{13.8}$$

To see this, check first that for $|S| = h$,

$$\tau_1(K_h, \{E\}, V) = n^{-1/(h-1)}m^{-1/2}.$$

Let $S \subset V$, with $s = |S| \leq h-1$, and consider restricted clique covers with cliques of size at most two, and exactly two.

For τ_1, the clique cover restricted to S is the edge-clique cover of K_s, plus a one-clique for each of the $h-s$ external edges for each vertex of K_s, so

$$\begin{aligned}\tau_1(K_h, \{E\}, S) &= \left(n^{s/[s(s-1)+s(h-s)]}m^{[s(s-1)/2+s(h-s)]/[s(s-1)+s(h-s)]}\right)^{-1} \\ &= \left(n^{1/(h-1)}m^{[h-(s+1)/2]/(h-1)}\right)^{-1} \\ &\leq \left(n^{1/(h-1)}m^{h/(2(h-1))}\right)^{-1} \\ &< \left(n^{1/(h-1)}m^{1/2}\right)^{-1},\end{aligned}$$

while for τ_2 we have

$$\tau_2(K_h, \{E\}, S) = \left(n^{1/(s-1)} m^{1/2}\right)^{-1} < \left(n^{1/(h-1)} m^{1/2}\right)^{-1},$$

thus verifying (13.8).

Let C be any irreducible clique cover of K_h (hence each clique has size at least two). We will show that for any fixed α,

$$\tau_1(K_h, C, V) \geq \begin{cases} \tau_1(K_h, \{V\}, V) & \text{for } \alpha \leq 2h/(h-1), \\ \tau_1(K_h, \{E\}, V) & \text{for } \alpha \geq 2h/(h-1). \end{cases}$$

Thus,

$$\tau_1(K_h, C, V) \geq \min\left\{\tau_1(K_h, \{V\}, V), \tau_1(K_h, \{E\}, V)\right\}. \tag{13.9}$$

Because $m = n^{\alpha}$, we see that

$$\tau_1(K_h, C, V) = n^{-x_C(\alpha)},$$

where

$$x_C(\alpha) = \frac{h}{\sum C} + \frac{|C|}{\sum C}\alpha, \quad x_{\{V\}}(\alpha) = 1 + \frac{\alpha}{h}, \quad x_{\{E\}}(\alpha) = \frac{1}{h-1} + \frac{\alpha}{2}.$$

(To simplify notation, in the following equation, we have replaced $x_{\{V\}}, x_{\{E\}}$ by x_V, x_E, respectively.) Note that for $\alpha_0 = 2h/(h-1)$ exponents

$$x_V(\alpha_0) = x_E(\alpha_0) = 1 + \frac{2}{h-1}.$$

Moreover, for all values of $\alpha < \alpha_0$, the function $x_V(\alpha) > x_E(\alpha)$, while for $\alpha > \alpha_0$, the function $x_V(\alpha) < x_E(\alpha)$.

Now, observe that $x_C(0) = \frac{h}{\sum C} \leq 1$ since each vertex is in at least one clique of C. Hence $x_C(0) \leq x_V(0) = 1$. We will also show that $x_C(\alpha) \leq x_V(\alpha)$ for $\alpha > 0$. To see this, we need to bound $|C|/\sum C$.

Suppose that $u \in V(K_h)$ appears in the fewest number of cliques of C, and let r be the number of cliques $C_i \in C$ to which u belongs. Then

$$\sum C = \sum_{i:C_i \ni u} |C_i| + \sum_{i:C_i \not\ni u} |C_i| \geq ((h-1) + r) + 2(|C| - r),$$

where $h-1$ counts all other vertices aside from u since they must appear in some clique with u.

For any $v \in V(K_h)$, we have

$$\begin{aligned} \sum C + |\{i : C_i \ni v\}| - (h-1) &\geq \sum C + r - (h-1) \\ &\geq (h-1) + r + 2(|C| - r) + r - (h-1) \\ &= 2|C|. \end{aligned}$$

Summing the above-mentioned inequality over all $v \in V(K_h)$,

$$h \sum C + \sum C - h(h-1) \geq 2h|C|,$$

and dividing both sides by $2h \sum C$, we finally get

$$\frac{|C|}{\sum C} \leq \frac{h+1}{2h} - \frac{h-1}{2\sum C}.$$

Now, using the above-mentioned bound,

$$\begin{aligned} x_C(\alpha_0) &= \frac{h}{\sum C} + \frac{|C|}{\sum C}\left(\frac{2h}{h-1}\right) \\ &\leq \frac{h}{\sum C} + \left(\frac{h+1}{2h} - \frac{h-1}{2\sum C}\right)\left(\frac{2h}{h-1}\right) \\ &= 1 + \frac{2}{h-1} \\ &= x_V(\alpha_0). \end{aligned}$$

Now, since $x_C(\alpha) \leq x_V(\alpha)$ at both $\alpha = 0$ and $\alpha = \alpha_0$, and both functions are linear, $x_C(\alpha) \leq x_V(\alpha)$ throughout the interval $(0, \alpha_0)$.

Since $x_E(\alpha_0) = x_V(\alpha_0)$, we also have $x_C(\alpha_0) \leq x_E(\alpha_0)$. The slope of $x_C(\alpha)$ is $\frac{|C|}{\sum C}$, and by the assumption that C consists of cliques of size at least two, this is at most 1/2. But the slope of $x_E(\alpha)$ is exactly 1/2. Thus, for all $\alpha \geq \alpha_0$, $x_C(\alpha) \leq x_E(\alpha)$. Hence the bounds given by formula (13.9) hold.

One can show (see [105]) that for any irreducible clique-cover C that is not $\{V\}$ nor $\{E\}$,

$$\max_S\{\tau_1(K_h, C, S), \tau_2(K_h, C, S)\} \geq \tau_1(K_h, C, V).$$

Hence, by (13.9),

$$\max_S\{\tau_1(K_h, C, S), \tau_2(K_h, C, S)\} \geq \min\{\tau_1(K_h, \{V\}, V), \tau_1(K_h, \{E\}, V)\}.$$

This implies that

$$\tau(K_h) = \begin{cases} n^{-1}m^{-1/h} & \text{for } \alpha \leq \alpha_0, \\ n^{-1/(h-1)}m^{-1/2} & \text{for } \alpha \geq \alpha_0, \end{cases}$$

which completes the proof of Corollary 13.6. □

To add to the picture of asymptotic behavior of small cliques in $G(n, m, p)$, we will quote the result of Rybarczyk and Stark [105], who obtained an upper bound on the total variation distance between the distribution of the number of h-cliques and a respective Poisson distribution for any fixed h.

Theorem 13.7 *Let $G(n, m, p)$ be a random intersection graph, where $m = n^\alpha$. Let $c > 0$ be a constant and $h \geq 3$ a fixed integer, and X_n be the random variable counting the number of copies of a complete graph K_h in $G(n, m, p)$.*

(i) If $\alpha < \frac{2h}{h-1}, p \sim cn^{-1}m^{-1/h}$, then

$$\lambda_n = \mathbb{E}X_n \sim c^h/h!$$

and

$$d_{TV}(\mathcal{L}(X_n), \text{Po}(\lambda_n)) = O\left(n^{-\alpha/h}\right).$$

(ii) If $\alpha = \frac{2h}{h-1}$, $p \sim cn^{-(h+1)/(h-1)}$, *then*

$$\lambda_n = \mathbb{E}X_n \sim \left(c^h + c^{h(h-1)}\right)/h!$$

and

$$d_{TV}(\mathcal{L}(X_n), \text{Po}(\lambda_n)) = O\left(n^{-2/(h-1)}\right).$$

(iii) If $\alpha > \frac{2h}{h-1}$, $p \sim cn^{-1/(h-1)}m^{-1/2}$, *then*

$$\lambda_n = \mathbb{E}X_n \sim c^{h(h-1)}/h!$$

and

$$d_{TV}(\mathcal{L}(X_n), \text{Po}(\lambda_n)) = O\left(n^{\left(h-\frac{\alpha(h-1)}{2}-\frac{2}{h-1}\right)} + n^{-1}\right).$$

Exercises

13.1.1 Show that if $p = \omega(n)/(n\sqrt{m})$, and $\omega(n) \to \infty$, then $G(n, m, p)$ has w.h.p. at least one edge.

13.1.2 Show that if $p = (2\log n + \omega(n))/m)^{1/2}$ and $\omega(n) \to -\infty$, then w.h.p. $G(n, m, p)$ is not complete.

13.1.3 Determine the expected number of isolated vertices in $G(m, n, p)$.

13.1.4 Draw all clique covers of a triangle.

13.2 Random Geometric Graphs

The graphs we consider in this section are the intersection graphs that we obtain from the intersections of balls in the d-dimensional unit cube, $D = [0, 1]^d$, where $d \geq 2$. For simplicity, we will only consider $d = 2$ in the text.

We let $\mathcal{X} = \{X_1, X_2, \ldots, X_n\}$ be independently and uniformly chosen from $D = [0, 1]^2$. For $r = r(n)$, let $G_{\mathcal{X},r}$ be the graph with vertex set $\mathcal{X}$. We join X_i, X_j by an edge if and only if X_j lies in the disk

$$B(X_i, r) = \left\{X \in [0, 1]^2 : |X - X_i| \leq r\right\}.$$

Here $|\ |$ denotes Euclidean distance.

For a given set $\mathcal{X}$ we see that increasing r can only add edges and so thresholds are usually expressed in terms of upper/lower bounds on the size of r.

The book by Penrose [100] gives a detailed exposition of this model. Our aim here is to prove some simple results that are not intended to be best possible.

Connectivity

The threshold (in terms of r) for connectivity was shown to be identical with that for minimum degree one, by Gupta and Kumar [57]. This was extended to k-connectivity by Penrose [99]. We do not aim for tremendous accuracy. The simple proof of connectivity was provided to us by Tobias Müller [92].

Theorem 13.8 *Let $\varepsilon > 0$ be arbitrarily small and let $r_0 = r_0(n) = \sqrt{\frac{\log n}{\pi n}}$. Then w.h.p.*

$$G_{X,r} \text{ contains isolated vertices if } r \leq (1-\varepsilon)r_0, \tag{13.10}$$

$$G_{X,r} \text{ is connected if } r \geq (1+\varepsilon)r_0. \tag{13.11}$$

Proof First consider (13.10) and the degree of X_1. Then

$$\mathbb{P}(X_1 \text{ is isolated}) \geq (1-\pi r^2)^{n-1}.$$

The factor $(1-\pi r^2)^{n-1}$ bounds the probability that none of $X_2, X_3, \ldots, X_n$ lie in $B(X_1, r)$ given that $B(X_1, r) \subseteq D$. It is exact for points far enough from the boundary of D.

Now,

$$(1-\pi r^2)^{n-1} \geq \left(1 - \frac{(1-\varepsilon)\log n}{n}\right)^n = n^{\varepsilon-1+o(1)}.$$

So if I is the set of isolated vertices, then $\mathbb{E}(|I|) \geq n^{\varepsilon-1+o(1)} \to \infty$. Now,

$$\mathbb{P}(X_1 \in I \mid X_2 \in I) \leq \left(1 - \frac{\pi r^2}{1-\pi r^2}\right)^{n-2} \leq (1+o(1))\,\mathbb{P}(X_1 \in I).$$

The expression $\left(1 - \frac{\pi r^2}{1-\pi r^2}\right)$ is the probability that a random point does not lie in $B(X_1, r)$ given that it does not lie in $B(X_2, r)$, and that $|X_2 - X_1| \geq 2r$. Equation (13.10) now follows from the Chebyshev inequality (2.18).

Now consider (13.11). Let $\eta \ll \varepsilon$ be a sufficiently small constant and divide D into ℓ_0^2 sub-squares of side length ηr, where $\ell_0 = 1/\eta r$. We refer to these sub-squares as cells. We can assume that η is chosen so that ℓ_0 is an integer. We say that a cell is *good* if contains at least $i_0 = \eta^3 \log n$ members of $\mathcal{X}$ and *bad* otherwise. We next let $K = 100/\eta^2$ and consider the number of bad cells in a $K \times K$ square block of cells.

Lemma 13.9 *Let B be a $K \times K$ square block of cells. The following hold w.h.p.:*

(a) If B is further than $100r$ from the closest boundary edge of D, then B contains at most $k_0 = (1-\varepsilon/10)\pi/\eta^2$ bad cells.

(b) If B is within distance $100r$ of exactly one boundary edge of D, then B contains at most $k_0/2$ bad cells.

(c) If B is within distance $100r$ of two boundary edges of D, then B contains no bad cells.

Proof (a) There are less than $\ell_0^2 < n$ such blocks. Furthermore, the probability that a fixed block contains k_0 or more bad cells is at most

$$\binom{K^2}{k_0}\left(\sum_{i=0}^{i_0}\binom{n}{i}(\eta^2r^2)^i(1-\eta^2r^2)^{n-i}\right)^{k_0}$$
$$\leq \left(\frac{K^2e}{k_0}\right)^{k_0}\left(2\left(\frac{ne}{i_0}\right)^{i_0}(\eta^2r^2)^{i_0}e^{-\eta^2r^2(n-i_0)}\right)^{k_0}. \quad (13.12)$$

The left-hand side (LHS) of (13.12) does require validation. We do this in terms of classic results about *balls in bins.* Think of the cells as M bins and the points $\mathcal{X}$ as balls. Then if W_S denotes the number of balls in the set of boxes S, then we have the following.

Lemma 13.10 *Let $S_1, S_2, \ldots, S_k$ be disjoint subsets of $[M]$ and let $s_1, s_2, \ldots, s_k$ be non-negative integers. Then*

$$\mathbb{P}\left(\bigcap_{i=1}^{k}\{W_{S_i} \leq s_i\}\right) \leq \prod_{i=1}^{k}\mathbb{P}(\{W_{S_i} \leq s_i\}),$$
$$\mathbb{P}\left(\bigcap_{i=1}^{k}\{W_{S_i} \geq s_i\}\right) \leq \prod_{i=1}^{k}\mathbb{P}(\{W_{S_i} \geq s_i\}).$$

The first inequality of Lemma 13.10 implies the LHS of (13.12). For a proof of Lemma 13.10, see, for example, Chapter 22.2. of [52]. Now,

$$\left(\frac{ne}{i_0}\right)^{i_0}(\eta^2r^2)^{i_0}e^{-\eta^2r^2(n-i_0)} \leq n^{-\eta^2(1+\varepsilon)/\pi} \quad (13.13)$$

for η sufficiently small. So we can bound the right-hand side (RHS) of (13.12) by

$$\left(\frac{2K^2en^{-\eta^2(1+\varepsilon)/\pi}}{(1-\varepsilon/10)\pi/\eta^2}\right)^{(1-\varepsilon/10)\pi/\eta^2} \leq n^{-1-\varepsilon/3}. \quad (13.14)$$

Part (a) follows after inflating the RHS of (13.14) by n to account for the number of choices of block.

(b) Replacing k_0 by $k_0/2$ replaces the LHS of (13.14) by

$$\left(\frac{4K^2en^{-\eta^2(1+\varepsilon/2)/\pi}}{(1-\varepsilon/10)\pi/2\eta^2}\right)^{(1-\varepsilon/10)\pi/2\eta^2} \leq n^{-1/2-\varepsilon/6}. \quad (13.15)$$

Observe now that the number of choices of block is $O(\ell_0) = o(n^{1/2})$ and then Part (b) follows after inflating the RHS of (13.15) by $o(n^{1/2})$ to account for the number of choices of block.

(c) Equation (13.13) bounds the probability that a single cell is bad. The number of cells in question in this case is $O(1)$ and (c) follows. □

We now do a simple geometric computation in order to place a lower bound on the number of cells within a ball $B(X,r)$.

Lemma 13.11 *A half-disk of radius $r_1 = r(1-\eta\sqrt{2})$ with diameter part of the grid of cells contains at least $(1-2\eta^{1/2})\pi/2\eta^2$ cells.*

Proof We place the half-disk in a $2r_1 \times r_1$ rectangle. Then we partition the rectangle into $\zeta_1 = r_1/r\eta$ rows of $2\zeta_1$ cells. The circumference of the circle will cut the ith row at a point which is $r_1(1-i^2\eta^2)^{1/2}$ from the center of the row. Thus, the ith row will contain at least $2\left\lfloor r_1(1-i^2\eta^2)^{1/2}/r\eta\right\rfloor$ complete cells. So the half-disk contains at least

$$\begin{aligned}\frac{2r_1}{r\eta}\sum_{i=1}^{1/\eta}((1-i^2\eta^2)^{1/2}-\eta) &\geq \frac{2r_1}{r\eta}\int_{x=1}^{1/\eta-1}((1-x^2\eta^2)^{1/2}-\eta)dx \\ &= \frac{2r_1}{r\eta^2}\int_{\theta=\arcsin(\eta)}^{\arcsin(1-\eta)}(\cos^2(\theta)-\eta\cos(\theta))d\theta \\ &\geq \frac{2r_1}{r\eta^2}\left[\frac{\theta}{2}-\frac{\sin(2\theta)}{4}-\eta\right]_{\theta=\arcsin(\eta)}^{\arcsin(1-\eta)}.\end{aligned}$$

Now,

$$\arcsin(1-\eta) \geq \frac{\pi}{2} - 2\eta^{1/2} \text{ and } \arcsin(\eta) \leq 2\eta.$$

So the number of cells is at least

$$\frac{2r_1}{r\eta^2}\left(\frac{\pi}{4}-\eta^{1/2}-\eta\right).$$

This completes the proof of Lemma 13.11. □

We deduce from Lemma 13.9 and Lemma 13.11 that

$$X \in \mathcal{X} \text{ implies that } B(X,r_1) \cap D \text{ contains at least one good cell.} \tag{13.16}$$

Now let Γ be the graph whose vertex set consists of the good cells and where cells c_1, c_2 are adjacent if and only if their centers are within distance r_1. Note that if c_1, c_2 are adjacent in Γ, then any point in $\mathcal{X} \cap c_1$ is adjacent in $G_{\mathcal{X},r}$ to any point in $\mathcal{X} \cap c_2$. It follows from (13.16) that all we need to do now is show that Γ is connected.

It follows from Lemma 13.9 that at most π/η^2 rows of a $K \times K$ block contain a bad cell. Thus, more than 95% of the rows and of the columns of such a block are free of bad cells. Call such a row or column good. The cells in a good row or column of some $K \times K$ block form part of the same component of Γ. Two neighboring blocks must have two touching good rows or columns so that the cells in a good row or column of some block form part of a single component of Γ. Any other component C must be in a block bounded by good rows and columns. But the existence of such a component means that it is surrounded by bad cells and then by Lemma 13.11 that there is a block B with at least $(1-3\eta^{1/2})\pi/\eta^2$ bad cells if it is far from the boundary and at least half of this if it is close to the boundary. But this contradicts Lemma 13.9. To see this, consider a cell in C whose center c has the largest second component, i.e., is highest in C. Now consider the half-disk H of radius r_1 that is centered at c. We can assume (i) H is contained entirely in B and (ii) at least $(1-2\eta^{1/2})\pi/2\eta^2 - (1-\eta\sqrt{2})/\eta \geq (1-3\eta^{1/2})\pi/2\eta^2$ cells in H are bad. Property (i) arises because cells above c whose centers are at distance at most r_1 are all bad and for (ii) we have discounted any bad cells on the diameter through c that might be in C. This provides half the claimed bad cells. We obtain the rest by considering a lowest cell of C. Near the boundary, we only need to consider one half-disk with diameter parallel to the closest boundary. Finally, observe that there are no bad cells close to a corner. □

Hamiltonicity

The first inroads on the Hamilton cycle problem were made by Diaz, Mitsche, and Pérez-Giménez [38]. Best possible results were later given by Balogh, Bollobás, Krivelevich, Müller, and Walters [12] and by Müller, Pérez, and Wormald [91]. As one might expect Hamiltonicity has a threshold at r close to r_0, as defined in Theorem 13.8. We now have enough to prove the result from [38]. We start with a simple lemma, taken from [12].

Lemma 13.12 *The subgraph Γ contains a spanning tree of maximum degree at most six.*

Proof Consider a spanning tree T of Γ that minimizes the sum of the lengths of the edges joining the centers of the cells. Then T does not have any vertex of degree greater than six. This is because, if center v were to have degree at least seven, then there are two neighboring centers u, w of v such that the angle between the line segments $[v, u]$ and $[v, w]$ is strictly less than 60 degrees. We can assume without loss of generality that $[v, u]$ is shorter than $[v, w]$. Note that if we remove the edge $\{v, w\}$ and add the edge $\{u, w\}$, then we obtain another spanning tree but with strictly smaller total edge length, a contradiction. Hence, T has maximum degree at most six. □

Theorem 13.13 *Suppose that $r \geq (1+\varepsilon)r_0$. Then w.h.p. $G_{\mathcal{X},r}$ is Hamiltonian.*

Proof We begin with the tree T promised by Lemma 13.12. Let c be a good cell. We partition the points of $\mathcal{X} \cap c$ into $2d$ roughly equal size sets $P_1, P_2, \ldots, P_{2d}$, where $d \leq 6$ is the degree of c in T. Since the points of $\mathcal{X} \cap c$ form a clique in $G = G_{\mathcal{X},r}$, we can form $2d$ paths in G from this partition.

We next do a walk W through T, for example, by breadth-first search that goes through each edge of T twice and passes through each node of Γ a number of times equal to twice its degree in Γ. Each time we pass through a node we traverse the vertices of a new path described in the previous paragraph. In this way, we create a cycle H that goes through all the points in $\mathcal{X}$ that lie in good cells.

Now, consider the points P in a bad cell c with center x. We create a path in G through P with endpoints x, y, say. Now, choose a good cell c' contained in the ball

$B(x, r_1)$ and then choose an edge $\{u, v\}$ of H in the cell c'. We merge the points in P into H by deleting $\{u, v\}$ and adding $\{x, u\}, \{y, v\}$. To make this work, we must be careful to ensure that we only use an edge of H at most once. But there are $\Omega(\log n)$ edges of H in each good cell and there are $O(1)$ bad cells within distance $2r$ say of any good cell and so this is easily done. □

Chromatic Number

We look at the chromatic number of $G_{\mathcal{X},r}$ in a limited range. Suppose that $n\pi r^2 = \frac{\log n}{\omega_r}$, where $\omega_r \to \infty, \omega_r = O(\log n)$. We are below the threshold for connectivity here. We will show that w.h.p.

$$\chi(G_{\mathcal{X},r}) \sim \Delta(G_{\mathcal{X},r}) \sim cl(G_{\mathcal{X},r}),$$

where we will use cl to denote the size of the largest clique. This is a special case of a result of McDiarmid [87]. We first bound the maximum degree.

Lemma 13.14

$$\Delta(G_{\mathcal{X},r}) \sim \frac{\log n}{\log \omega_r} \quad w.h.p.$$

Proof Let Z_k denote the number of vertices of degree k and let $Z_{\geq k}$ denote the number of vertices of degree at least k. Let $k_0 = \frac{\log n}{\omega_d}$, where $\omega_d \to \infty$ and $\omega_d = o(\omega_r)$. Then

$$\mathbb{E}(Z_{\geq k_0}) \leq n\binom{n}{k_0}(\pi r^2)^{k_0} \leq n\left(\frac{ne\omega_d \log n}{n\omega_r \log n}\right)^{\frac{\log n}{\omega_d}} = n\left(\frac{e\omega_d}{\omega_r}\right)^{\frac{\log n}{\omega_d}}.$$

So,

$$\log(\mathbb{E}(Z_{\geq k_0})) \leq \frac{\log n}{\omega_d}\left(\omega_d + 1 + \log \omega_d - \log \omega_r\right). \tag{13.17}$$

Now let $\varepsilon_0 = \omega_r^{-1/2}$. Then if

$$\omega_d + \log \omega_d + 1 \leq (1 - \varepsilon_0) \log \omega_r,$$

then (13.17) implies that $\mathbb{E}(Z_k) \to 0$. This verifies the upper bound on Δ claimed in the lemma.

Now let $k_1 = \frac{\log n}{\widehat{\omega}_d}$, where $\widehat{\omega}_d$ is the solution to

$$\widehat{\omega}_d + \log \widehat{\omega}_d + 1 = (1 + \varepsilon_0) \log \omega_r.$$

Next let M denote the set of vertices that are at distance greater than r from any edge of D. Let M_k be the set of vertices of degree k in M. If $\widehat{Z}_k = |M_k|$, then

$$\mathbb{E}(\widehat{Z}_{k_1}) \geq n\,\mathbb{P}(X_1 \in M) \times \binom{n-1}{k_1}(\pi r^2)^{k_1}(1 - \pi r^2)^{n-1-k_1}.$$

$\mathbb{P}(X_1 \in M) \geq 1 - 4r$ and so

$$\mathbb{E}(\widehat{Z}_{k_1}) \geq (1-4r)\frac{n}{3k_1^{1/2}}\left(\frac{(n-1)e}{k_1}\right)^{k_1}(\pi r^2)^{k_1}e^{-n\pi r^2/(1-\pi r^2)}$$

$$\geq (1-o(1))\frac{n^{1-1/\omega_r}}{3k_1^{1/2}}\left(\frac{e\widehat{\omega}_d}{\omega_r}\right)^{\frac{\log n}{\widehat{\omega}_d}}.$$

So,

$$\log(\mathbb{E}(\widehat{Z}_{k_1})) = \Omega\left(\frac{\log n}{\omega_r^{1/2}}\right) \to \infty. \tag{13.18}$$

An application of the Chebyshev inequality finishes the proof of the lemma. Indeed,

$$\mathbb{P}(X_1, X_2 \in M_k) \leq \mathbb{P}(X_1 \in M)\,\mathbb{P}(X_2 \in M)$$
$$\times\left(\mathbb{P}(X_2 \in B(X_1, r)) + \left(\binom{n-1}{k_1}(\pi r^2)^{k_1}(1-\pi r^2)^{n-2k_1-2}\right)^2\right)$$
$$\leq (1+o(1))\,\mathbb{P}(X_1 \in M_k)\,\mathbb{P}(X_2 \in M_k).$$

□

Now $\mathrm{cl}(G_{\mathcal{X},r}) \leq \Delta(G_{\mathcal{X},r}) + 1$ and so we now lower bound $\mathrm{cl}(G_{\mathcal{X},r})$ w.h.p. But this is easy. It follows from Lemma 13.14 that w.h.p. there is a vertex X_j with at least $(1-o(1))\frac{\log n}{\log(4\omega_r)}$ vertices in its $r/2$ ball $B(X_j, r/2)$. But such a ball provides a clique of size $(1-o(1))\frac{\log n}{\log(4\omega_r)}$. We have therefore proved the following.

Theorem 13.15 *Suppose that $n\pi r^2 = \frac{\log n}{\omega_r}$, where $\omega_r \to \infty$, $\omega_r = O(\log n)$. Then w.h.p.*

$$\chi(G_{\mathcal{X},r}) \sim \Delta(G_{\mathcal{X},r}) \sim cl(G_{\mathcal{X},r}) \sim \frac{\log n}{\log \omega_r}.$$

We now consider larger r.

Theorem 13.16 *Suppose that $n\pi r^2 = \omega_r \log n$, where $\omega_r \to \infty$, $\omega_r = o(n/\log n)$. Then w.h.p.*

$$\chi(G_{\mathcal{X},r}) \sim \frac{\omega_r\sqrt{3}\log n}{2\pi}.$$

Proof First consider the triangular lattice in the plane. This is the set of points $T = \{m_1 a + m_2 b : m_1, m_2 \in \mathbb{Z}\}$, where $a = (0,1), b = (1/2, \sqrt{3}/2)$ (see Figure 13.1). As in the diagram, each $v \in T$ can be placed at the center of a hexagon C_v. The C_v intersect on a set of measure zero and each C_v has area $\sqrt{3}/2$ and is contained in $B(v, 1/\sqrt{3})$. Let $\Gamma(T,d)$ be the graph with vertex set T, where two vertices $x, y \in T$ are joined by an edge if their Euclidean distance $|x-y| < d$.

Lemma 13.17 (McDiarmid and Reed [88])

$$\chi(\Gamma(T,d)) \leq (d+1)^2.$$

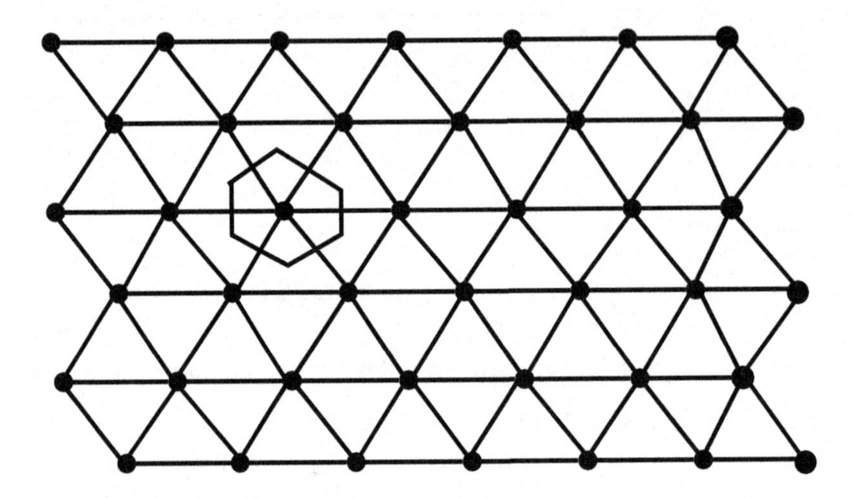

Figure 13.1 The small hexagon is an example of a C_v

Proof Let $\delta = \lceil d \rceil$. Let R denote a $\delta \times \delta$ rhombus made up of triangles of T with one vertex at the origin. This rhombus has δ^2 vertices if we exclude those at the top and right-hand end. We give each of these vertices a distinct color and then tile the plane with copies of R. This is a proper coloring, by construction. □

Armed with this lemma we can easily get an upper bound on $\chi(G_{\mathcal{X},r})$. Let $\delta = 1/\omega_r^{1/3}$ and let $s = \delta r$. Let sT be the contraction of the lattice T by a factor s, i.e., $sT = \{sx : x \in T\}$. Then if $v \in sT$, let sC_v be the hexagon with center v, sides parallel to the sides of C_v but reduced by a factor s. $|\mathcal{X} \cap sC_v|$ is distributed as $\text{Bin}(n, s^2\sqrt{3}/2)$. So the Chernoff bounds imply that with probability $1 - o(n^{-1})$,

$$sC_v \text{ contains} \leq \theta = \left\lceil (1 + \omega_r^{-1/8}) ns^2\sqrt{3}/2 \right\rceil \text{ members of } \mathcal{X}. \tag{13.19}$$

Let $\rho = r + 2s/\sqrt{3}$. We note that if $x \in C_v$ and $y \in C_w$ and $|x - y| \leq r$, then $|v - w| \leq \rho$. Thus, given a proper coloring φ of $\Gamma(sT, \rho)$ with colors $[q]$ we can w.h.p. extend it to a coloring ψ of $G_{\mathcal{X},r}$ with color's $[q] \times [\theta]$. If $x \in sC_v$ and $\varphi(x) = a$, then we let $\psi(x) = (a, b)$, where b ranges over $[\theta]$ as x ranges over $sC_v \cap \mathcal{X}$. So, w.h.p.

$$\chi(G_{\mathcal{X},r}) \leq \theta\chi(\Gamma(sT, \rho)) = \theta\chi(\Gamma(T, \rho/s)) \leq \theta\left(\frac{\rho}{s} + 1\right)^2$$

$$\sim \frac{ns^2\sqrt{3}}{2} \times \frac{r^2}{s^2} = \frac{\omega_r\sqrt{3}\log n}{2\pi}. \tag{13.20}$$

For the lower bound, we use a classic result on packing disks in the plane.

Lemma 13.18 *Let $A_n = [0, n]^2$ and C be a collection of disjoint disks of unit area that touch A_n. Then $|C| \leq (1 + o(1))\pi n^2/\sqrt{12}$.*

Proof Thue's theorem states that the densest packing of disjoint same size disks in the plane is the hexagonal packing which has density $\lambda = \pi/\sqrt{12}$. Let C' denote the disks that are contained entirely in A_n. Then we have

$$|C'| \geq |C| - O(n) \text{ and } |C'| \leq \frac{\pi n^2}{\sqrt{12}}.$$

The first inequality comes from the fact that if $C \in C \setminus C'$, then it is contained in a perimeter of width $O(1)$ surrounding A_n. □

Now consider the subgraph H of $G_{\mathcal{X},r}$ induced by the points of $\mathcal{X}$ that belong to the square with center $(1/2, 1/2)$ and side $1-2r$. It follows from Lemma 13.18 that if $\alpha(H)$ is the size of the largest independent set in H, then $\alpha(H) \leq (1+o(1))2/(r^2\sqrt{3})$. This is because if S is an independent set of H, then the disks $B(x, r/2)$ for $x \in S$ are necessarily disjoint. Now using the Chernoff bounds, we see that w.h.p. H contains at least $(1-o(1))n$ vertices. Thus,

$$\chi(G_{\mathcal{X},r}) \geq \chi(H) \geq \frac{|V(H)|}{\alpha(H)} \geq (1-o(1))\frac{r^2\sqrt{3}n}{2} = (1-o(1))\frac{\omega_r\sqrt{3}\log n}{2\pi}.$$

This completes the proof of Theorem 13.16. □

Exercises

13.2.1 Verify equation (13.13).

13.2.2 Verify equation (13.18).

Problems for Chapter 13

13.1 Prove that the bound (13.2) holds.

13.2 Prove that the bound (13.3) holds.

13.3 Prove that the bound (13.4) holds.

13.4 Prove the claims in Lemma 13.2.

13.5 Let X denote the number of isolated vertices in the binomial random intersection graph $G(n, m, p)$, where $m = n^\alpha$, $\alpha > 0$. Show that if

$$p = \begin{cases} (\log n + \varphi(n))/m & \text{when } \alpha \leq 1, \\ \sqrt{(\log n + \varphi(n))/(nm)} & \text{when } \alpha > 1, \end{cases}$$

then $\mathbb{E}\, X \to e^{-c}$ if $\lim_{n\to\infty} \varphi(n) \to c$ for any real c.

13.6 Find the variance of the random variable X counting isolated vertices in $G(n, m, p)$.

13.7 Let Y be a random variable that counts vertices of degree greater than one in $G(n, m, p)$, with $m = n^\alpha$ and $\alpha > 1$. Show that for $p^2m^2n \gg \log n$

$$\lim_{n\to\infty} \mathbb{P}\left(Y > 2p^2m^2n\right) = 0.$$

13.8 Suppose that $r \geq (1+\varepsilon)r_0$, as in Theorem 13.8. Show that if $1 \leq k = O(1)$, then $G_{\mathcal{X},r}$ is k-connected w.h.p.

13.9 Show that if $2 \leq k = O(1)$ and $r \gg n^{-\frac{k}{2(k-1)}}$, then w.h.p. $G_{\mathcal{X},r}$ contains a k-clique. On the other hand, show that if $r = o(n^{-\frac{k}{2(k-1)}})$, then $G_{\mathcal{X},r}$ contains no k-clique.

13.10 Suppose that $r \gg \sqrt{\frac{\log n}{n}}$. Show that w.h.p. the diameter of $G_{\mathcal{X},r} = \Theta\left(\frac{1}{r}\right)$.

13.11 Suppose that $r \geq (1+\varepsilon)r_0$, as in Theorem 13.8. Show that if $2 \leq k = O(1)$, then $G_{\mathcal{X},r}$ has k edge disjoint Hamilton cycles w.h.p.

13.12 Given $\mathcal{X}$ and an integer k we define the k-nearest neighbor graph $G_{k-NN,\mathcal{X}}$ as follows: We add an edge between x and y of $\mathcal{X}$ if and only if y is one of x's k-nearest neighbors, in Euclidean distance or vice versa. Show that if $k \geq C \log n$ for a sufficiently large C, then $G_{k-NN,\mathcal{X}}$ is connected w.h.p.

13.13 Suppose that we independently deposit n random black points $\mathcal{X}_b$ and n random white points $\mathcal{X}_w$ into D. Let $B_{\mathcal{X}_b,\mathcal{X}_w,r}$ be the bipartite graph where we connect $x \in \mathcal{X}_b$ with $\mathcal{X}_w$ if and only if $|x-y| \leq r$. Show that if $r \gg \sqrt{\frac{\log n}{n}}$, then w.h.p. $B_{\mathcal{X}_b,\mathcal{X}_w,r}$ contains a perfect matching.

14 Weighted Graphs

There are many cases in which we put weights $X_e, e \in E$ on the edges of a graph or digraph and ask for the minimum or maximum weight object. The optimization questions that arise from this are the backbone of combinatorial optimization. When the X_e are random variables, we can ask for properties of the optimum value, which will be a random variable. In this chapter, we consider three of the most basic optimization problems:

Minimum weight spanning trees. Suppose one wishes to connect a set of n points in the plane together by lines as cheaply as possible so that the graph formed is connected. The solution is to construct a minimum length spanning tree. Analysis of the famous greedy algorithm leads to an elegant estimate of the expected minimum length when edge lengths are independent uniform $[0, 1]$.

Shortest paths. This is the classic problem of finding the best route through a network. This is the problem faced by any driver trying to get from A to B as quickly as possible. Analysis of Dijkstra's famous shortest path algorithm when edge lengths are independent exponential mean one together with the memoryless property of this distribution leads to very precise estimates of expected shortest path lengths.

Minimum weight matchings in bipartite graphs. This is the problem faced by someone who has n tasks to fulfill and n persons who can fill each task with a different level of competence. Or any circumstance where one has to assign n objects to n positions where there is a cost $C(i, j)$ of assigning object i to position j. When edge costs are independent exponential mean one, we arrive at one of the most striking answers to estimating expectation one can conceive.

14.1 Minimum Weight Spanning Tree

Let the edges of the complete graph K_n on $[n]$ be given independent lengths X_e, $e \in \binom{[n]}{2}$. Here X_e is a uniform $[0, 1]$ random variable. Let L_n be the length of the minimum spanning tree (MST) of K_n with these edge lengths.

Frieze [50] proved the following theorem. The proof we give utilizes the rather lovely integral formula (14.1) due to Janson [62] (see also the related equation (7) from [53]).

Theorem 14.1

$$\lim_{n\to\infty} \mathbb{E}\, L_n = \zeta(3) = \sum_{k=1}^{\infty} \frac{1}{k^3} = 1.202\ldots$$

Proof Suppose that $T = T(\{X_e\})$ is the MST, unique with probability one. We use the identity

$$a = \int_0^1 1_{\{x\le a\}} dx.$$

Therefore,

$$\begin{aligned} L_n &= \sum_{e\in T} X_e \\ &= \sum_{e\in T} \int_{p=0}^1 1_{\{p\le X_e\}} dp \\ &= \int_{p=0}^1 \sum_{e\in T} 1_{\{p\le X_e\}} dp \\ &= \int_{p=0}^1 |\{e \in T : X_e \ge p\}| dp \\ &= \int_{p=0}^1 (\kappa(\mathbb{G}_p) - 1) dp, \end{aligned}$$

where $\kappa(\mathbb{G}_p)$ denotes the number of components of graph $\mathbb{G}_p$. Here $\mathbb{G}_p$ is the graph induced by the edges e with $X_e \le p$, i.e., $\mathbb{G}_p \equiv \mathbb{G}_{n,p}$. The last line may be considered to be a consequence of the fact that the greedy algorithm solves the MST problem. This algorithm examines edges in increasing order of edge weight. It builds a tree, adding one edge at a time. It adds the edge to the forest F of edges accepted so far, only if the two endpoints lie in distinct components of F. Otherwise, it moves onto the next edge. Thus, the number of edges to be added given F is $\kappa(F) - 1$ and if the longest edge in $e \in F$ has $X_e = p$, then $\kappa(F) = \kappa(\mathbb{G}_p)$, which follows by an easy induction. Hence

$$\mathbb{E}\, L_n = \int_{p=0}^1 (\mathbb{E}\, \kappa(\mathbb{G}_p) - 1) dp. \tag{14.1}$$

We therefore estimate $\mathbb{E}\,\kappa(\mathbb{G}_p)$. We observe first that

$$p \geq \frac{6\log n}{n} \Rightarrow \mathbb{E}\,\kappa(\mathbb{G}_p) = 1 + o(1). \tag{14.2}$$

Hence, if $p_0 = \frac{6\log n}{n}$, then

$$\begin{aligned}\mathbb{E}\,L_n &= \int_{p=0}^{p_0} (\mathbb{E}\,\kappa(\mathbb{G}_p) - 1)dp + o(1)\\ &= \int_{p=0}^{p_0} \mathbb{E}\,\kappa(\mathbb{G}_p)dp + o(1).\end{aligned}$$

Write

$$\kappa(\mathbb{G}_p) = \sum_{k=1}^{(\log n)^2} A_k + \sum_{k=1}^{(\log n)^2} B_k + C,$$

where A_k stands for the number of components which are k vertex trees, B_k is the number of k vertex components which are not trees, and, finally, C denotes the number of components on at least $(\log n)^2$ vertices. Then, for $1 \leq k \leq (\log n)^2$ and $p \leq p_0$,

$$\begin{aligned}\mathbb{E}\,A_k &= \binom{n}{k} k^{k-2} p^{k-1} (1-p)^{k(n-k)+\binom{k}{2}-k+1}\\ &= (1+o(1))n^k \frac{k^{k-2}}{k!} p^{k-1}(1-p)^{kn},\end{aligned}$$

$$\begin{aligned}\mathbb{E}\,B_k &\leq \binom{n}{k} k^{k-2} \binom{k}{2} p^k (1-p)^{k(n-k)}\\ &\leq (1+o(1))(npe^{1-np})^k\\ &\leq 1 + o(1),\end{aligned}$$

$$C \leq \frac{n}{(\log n)^2}.$$

Hence

$$\int_{p=0}^{\frac{6\log n}{n}} \sum_{k=1}^{(\log n)^2} \mathbb{E}\,B_k\, dp \leq \frac{6\log n}{n}(\log n)^2(1+o(1)) = o(1),$$

and

$$\int_{p=0}^{\frac{6\log n}{n}} C\, dp \leq \frac{6\log n}{n} \frac{n}{(\log n)^2} = o(1).$$

So

$$\mathbb{E}\,L_n = o(1) + (1+o(1)) \sum_{k=1}^{(\log n)^2} n^k \frac{k^{k-2}}{k!} \int_{p=0}^{\frac{6\log n}{n}} p^{k-1}(1-p)^{kn} dp.$$

But

$$\sum_{k=1}^{(\log n)^2} n^k \frac{k^{k-2}}{k!} \int_{p=\frac{6\log n}{n}}^{1} p^{k-1}(1-p)^{kn} dp$$
$$\leq \sum_{k=1}^{(\log n)^2} n^k \frac{k^{k-2}}{k!} \int_{p=\frac{6\log n}{n}}^{1} n^{-6k} dp$$
$$= o(1).$$

Therefore,

$$\mathbb{E} L_n = o(1) + (1 + o(1)) \sum_{k=1}^{(\log n)^2} n^k \frac{k^{k-2}}{k!} \int_{p=0}^{1} p^{k-1}(1-p)^{kn} dp$$
$$= o(1) + (1 + o(1)) \sum_{k=1}^{\infty} \frac{1}{k^3}. \tag{14.3}$$

□

One can obtain the same result if the uniform $[0, 1]$ random variable is replaced by any non-negative random variable with distribution F having a derivative equal to one at the origin, e.g., an exponential variable with mean one (see Steele [109]).

Exercises

14.1.1 Verify equation (14.2).
14.1.2 Verify equation (14.3).

14.2 Shortest Paths

Let the edges of the complete graph K_n on $[n]$ be given independent lengths X_e, $e \in \binom{[n]}{2}$. Here X_e is exponentially distributed with mean one. The following theorem was proved by Janson [63]:

Theorem 14.2 *Let X_{ij} be the distance from vertex i to vertex j in the complete graph with edge weights independent $EXP(1)$ random variables. Then, for every $\varepsilon > 0$, as $n \to \infty$,*

(i) for any fixed i, j,

$$\mathbb{P}\left(\left|\frac{X_{ij}}{\log n/n} - 1\right| \geq \varepsilon\right) \to 0;$$

(ii) for any fixed i,

$$\mathbb{P}\left(\left|\frac{\max_j X_{ij}}{\log n/n} - 2\right| \geq \varepsilon\right) \to 0;$$

(iii)

$$\mathbb{P}\left(\left|\frac{\max_{i,j} X_{ij}}{\log n/n} - 3\right| \geq \varepsilon\right) \to 0.$$

Proof First, recall the following two properties of the exponential X:

(P1) $\mathbb{P}(X > \alpha + \beta | X > \alpha) = \mathbb{P}(X > \beta)$.

(P2) If $X_1, X_2, \dots, X_m$ are independent $EXP(1)$ exponential random variables, then $\min\{X_1, X_2, \dots, X_m\}$ is an exponential with mean $1/m$.

Suppose that we want to find shortest paths from a vertex s to all other vertices in a digraph with non-negative arc lengths. Recall Dijkstra's algorithm. After several iterations, there is a rooted tree T such that if v is a vertex of T, then the tree path from s to v is a shortest path. Let $d(v)$ be its length. For $x \notin T$, let $d(x)$ be the minimum length of a path P that goes from s to v to x where $v \in T$ and the sub-path of P that goes to v is the tree path from s to v. If $d(y) = \min\{d(x) : x \notin T\}$, then $d(y)$ is the length of a shortest path from s to y and y can be added to the tree.

Suppose that vertices are added to the tree in the order $v_1, v_2, \dots, v_n$ and that $Y_j = \text{dist}(v_1, v_j)$ for $j = 1, 2, \dots, n$. It follows from property P1 that

$$Y_{k+1} = \min_{\substack{i=1,2,\dots,k \\ v \neq v_1,\dots,v_k}} [Y_i + X_{v_i,v}] = Y_k + E_k,$$

where E_k is exponential with mean $\frac{1}{k(n-k)}$ and is independent of Y_k.

This is because X_{v_i,v_j} is distributed as an independent exponential X conditioned on $X \geq Y_k - Y_i$. Hence

$$\begin{aligned}\mathbb{E} Y_n &= \sum_{k=1}^{n-1} \frac{1}{k(n-k)} = \frac{1}{n}\sum_{k=1}^{n-1}\left(\frac{1}{k} + \frac{1}{n-k}\right) = \frac{2}{n}\sum_{k=1}^{n-1}\frac{1}{k} \\ &= \frac{2\log n}{n} + O(n^{-1}).\end{aligned}$$

Also, from the independence of E_k, Y_k,

$$\text{Var}\, Y_n = \sum_{k=1}^{n-1} \text{Var}\, E_k = \sum_{k=1}^{n-1}\left(\frac{1}{k(n-k)}\right)^2 \leq 2\sum_{k=1}^{n/2}\left(\frac{1}{k(n-k)}\right)^2$$

$$\leq \frac{8}{n^2} \sum_{k=1}^{n/2} \frac{1}{k^2} = O(n^{-2})$$

and we can use the Chebyshev inequality (2.18) to prove (ii).

Now fix $j = 2$. Then, if i is defined by $v_i = 2$, we see that i is uniform over $\{2, 3, \ldots, n\}$. So

$$\mathbb{E}\, X_{1,2} = \frac{1}{n-1} \sum_{i=2}^{n} \sum_{k=1}^{i-1} \frac{1}{k(n-k)} = \frac{1}{n-1} \sum_{k=1}^{n-1} \frac{n-k}{k(n-k)}$$

$$= \frac{1}{n-1} \sum_{k=1}^{n-1} \frac{1}{k} = \frac{\log n}{n} + O(n^{-1}).$$

For the variance of $X_{1,2}$ we have

$$X_{1,2} = \delta_2 Y_2 + \delta_3 Y_3 + \cdots + \delta_n Y_n,$$

where

$$\delta_i \in \{0, 1\}; \quad \delta_2 + \delta_3 + \cdots + \delta_n = 1; \quad \mathbb{P}(\delta_i = 1) = \frac{1}{n-1}.$$

$$\operatorname{Var} X_{1,2} = \sum_{i=2}^{n} \operatorname{Var}(\delta_i Y_i) + \sum_{i \neq j} \operatorname{Cov}(\delta_i Y_i, \delta_j Y_j)$$

$$\leq \sum_{i=2}^{n} \operatorname{Var}(\delta_i Y_i).$$

The last inequality holds since

$$\operatorname{Cov}(\delta_i Y_i, \delta_j Y_j) = \mathbb{E}(\delta_i Y_i \delta_j Y_j) - \mathbb{E}(\delta_i Y_i)\, \mathbb{E}(\delta_j Y_j)$$

$$= -\mathbb{E}(\delta_i Y_i)\, \mathbb{E}(\delta_j Y_j) \leq 0.$$

So

$$\operatorname{Var} X_{1,2} \leq \sum_{i=2}^{n} \operatorname{Var}(\delta_i Y_i)$$

$$\leq \sum_{i=2}^{n} \frac{1}{n-1} \sum_{k=1}^{i-1} \left(\frac{1}{k(n-k)} \right)^2$$

$$= O(n^{-2}).$$

We can now use the Chebyshev inequality.

We turn now to proving (iii). We begin with a lower bound. Let $Y_i = \min \left\{ X_{i,j} : i \neq j \in [n] \right\}$. Let $A = \left\{ i : Y_i \geq \frac{(1-\varepsilon)\log n}{n} \right\}$. Then we have that for $i \in [n]$,

$$\mathbb{P}(i \in A) = \exp\left\{ -(n-1) \frac{(1-\varepsilon)\log n}{n} \right\} = n^{-1+\varepsilon+o(1)}. \tag{14.4}$$

An application of the Chebyshev inequality shows that $|A| \sim n^{\varepsilon+o(1)}$ w.h.p. Now the expected number of paths from $a_1 \in A$ to $a_2 \in A$ of length at most $\frac{(3-2\varepsilon)\log n}{n}$ can be bounded by

$$n^{2\varepsilon+o(1)} \times n^2 \times n^{-3\varepsilon+o(1)} \times \frac{\log^2 n}{n^2} = n^{-\varepsilon+o(1)}. \tag{14.5}$$

Explanation for (14.5): The first factor $n^{2\varepsilon+o(1)}$ is the expected number of pairs of vertices $a_1, a_2 \in A$. The second factor is a bound on the number of choices b_1, b_2 for the neighbors of a_1, a_2 on the path. The third factor F_3 is a bound on the expected number of paths of length at most $\frac{\alpha \log n}{n}$ from b_1 to b_2, $\alpha = 1 - 3\varepsilon$. This factor comes from

$$F_3 \leq \sum_{\ell \geq 0} n^\ell \left(\frac{\alpha \log n}{n}\right)^{\ell+1} \frac{1}{(\ell+1)!}.$$

Here ℓ is the number of internal vertices on the path. There will be at most n^ℓ choices for the sequence of vertices on the path. We then use the fact that the exponential mean one random variable stochastically dominates the uniform $[0,1]$ random variable U. The final two factors are the probability that the sum of $\ell + 1$ independent copies of U sum to at most $\frac{\alpha \log n}{n}$. Continuing, we have

$$F_3 \leq \sum_{\ell \geq 0} \frac{\alpha \log n}{n(\ell+1)} \left(\frac{e^{1+o(1)} \alpha \log n}{\ell}\right)^\ell = n^{-1+\alpha+o(1)}. \tag{14.6}$$

The final factor in (14.5) is a bound on the probability that $X_{a_1 b_1} + X_{a_2 b_2} \leq \frac{(2+\varepsilon)\log n}{n}$. For this, we use the fact that $X_{a_i b_i}, i = 1, 2$ is distributed as $\frac{(1-\varepsilon)\log n}{n} + E_i$, where E_1, E_2 are independent exponential mean one. Now $\mathbb{P}(E_1 + E_2 \leq t) \leq (1 - e^{-t})^2 \leq t^2$ and taking $t = \frac{3\varepsilon \log n}{n}$ justifies the final factor of (14.5).

It follows from (14.5) that w.h.p. the shortest distance between a pair of vertices in A is at least $\frac{(3-2\varepsilon)\log n}{n}$ w.h.p., completing our proof of the lower bound in (iii).

We now consider the upper bound. Let now $Y_1 = d_{k_3}$, where $k_3 = n^{1/2}\log n$. For $t < 1 - \frac{1+o(1)}{n}$ we have that

$$\mathbb{E}(e^{tnY_1}) = \mathbb{E}\left(\exp\left\{\sum_{i=1}^{k_3} \frac{tn}{i(n-i)}\right\}\right) = \prod_{i=1}^{k} \left(1 - \frac{(1+o(1))t}{i}\right)^{-1}.$$

Then for any $\alpha > 0$ we have

$$\Pr\left(Y_1 \geq \frac{\alpha \log n}{n}\right) \leq \mathbb{E}(e^{tnY_1 - t\alpha \log n}) = O\left(\exp\left\{\left(\frac{1}{2} + o(1) - \alpha\right) t \log n\right\}\right). \tag{14.7}$$

It follows, on taking $\alpha = 3/2 + o(1)$ that w.h.p.

$$Y_j \leq \frac{(3+o(1))\log n}{2n} \quad \text{for all } j \in [n].$$

Letting T_j be the set corresponding to S_{k_3} when we execute Dijkstra's algorithm starting at j, then we have that for $j \neq k$, where $T_j \cap T_k = \emptyset$,

$$\mathbb{P}\left(\nexists e \in E(T_j,T_k) : X_e \leq \frac{\log n}{n}\right) \leq \exp\left\{-\frac{k_3^2 \log n}{n}\right\}$$
$$= e^{-(2+o(1))\log^2 n} = o(n^{-2}),$$

and this is enough to complete the proof of (iii). □

We can, as for spanning trees, replace the exponential random variables by random variables that behave like the exponential close to the origin. The paper of Janson [63] allows for any random variable X satisfying $\mathbb{P}(X \leq t) = t + o(t)$ as $t \to 0$.

Exercises

14.2.1 Verify equation (14.6).

14.2.2 Verify equation (14.7).

14.3 Minimum Weight Assignment

Consider the complete bipartite graph $K_{n,n}$ and suppose that its edges are assigned independent exponentially distributed weights, with rate 1. (The rate of an exponential variable is one over its mean.) Denote the minimum total weight of a perfect matching in $K_{n,n}$ by C_n. Aldous [5, 6] proved that $\lim_{n\to\infty} \mathbb{E}\, C_n = \zeta(2) = \sum_{k=1}^{\infty} \frac{1}{k^2}$. The following theorem was conjectured by Parisi [98]. It was proved independently by Linusson and Wästlund [78] and Nair, Prabhakar, and Sharma [94]. The proof given here is from Wästlund [112].

Theorem 14.3

$$\mathbb{E}\, C_n = \sum_{k=1}^{n} \frac{1}{k^2} = 1 + \frac{1}{4} + \frac{1}{9} + \frac{1}{16} + \cdots + \frac{1}{n^2}. \tag{14.8}$$

From the above-mentioned theorem we immediately get the following corollary, first proved by Aldous [6].

Corollary 14.4

$$\lim_{n\to\infty} \mathbb{E}\, C_n = \zeta(2) = \sum_{k=1}^{\infty} \frac{1}{k^2} = \frac{\pi^2}{6} = 1.6449\ldots$$

Let $EXP(\lambda)$ denote an exponential random variable of rate λ, i.e., $\mathbb{P}(EXP(\lambda) \geq x) = e^{-\lambda x}$. Consider the complete bipartite graph $K_{n,n}$, with bipartition (A, B), where $A = \{a_1, a_2, \ldots, a_n\}$ and $B = \{b_1, b_2, \ldots, b_n\}$, and with edge weights that are independent copies of $EXP(1)$. We add a special vertex b^* to B, with edges to all n vertices of A. Each edge adjacent to b^* is assigned an $EXP(\lambda)$ weight independently, $\lambda > 0$.

For $r \geq 1$, we let M_r be the minimum weight matching of $A_r = \{a_1, a_2, \ldots, a_r\}$ into B and M_r^* be the minimum weight matching of A_r into $B^* = B \cup \{b^*\}$. (As $\lambda \to 0$, it becomes increasingly unlikely that any of the extra edges is actually used in the minimum weight matching.) We denote this matching by M_r^* and we let B_r^* denote the corresponding set of vertices of B^* that are covered by M_r^*. We let $C(n, r)$ denote the weight of M_r.

Define $P(n, r)$ as the normalized probability that b^* participates in M_r^*, i.e.,

$$P(n, r) = \lim_{\lambda\to 0} \frac{\mathbb{P}(b^* \in B_r^*)}{\lambda}. \tag{14.9}$$

Its importance lies in the following lemma:

Lemma 14.5

$$\mathbb{E}(C(n, r) - C(n, r - 1)) = \frac{P(n, r)}{r}. \tag{14.10}$$

Proof Choose i randomly from $[r]$ and let $\widehat{B}_i \subseteq B_r$ be the B-vertices in the minimum weight matching of $(A_r \setminus \{a_i\})$ into B^*. Let $X = C(n,r)$ and $Y = C(n, r-1)$. Let w_i be the weight of the edge (a_i, b^*) and I_i denote the indicator variable for the event that the minimum weight of an A_r matching that contains this edge is smaller than the minimum weight of an A_r matching that does not use b^*. We can see that I_i is the indicator variable for the event $\{Y_i + w_i < X\}$, where Y_i is the minimum weight of a matching from $A_r \setminus \{a_i\}$ to B. Indeed, if $(a_i, b^*) \in M_r^*$, then $w_i < X - Y_i$. Conversely, if $w_i < X - Y_i$ and no other edge from b^* has weight smaller than $X - Y_i$, then $(a_i, b^*) \in M_r^*$, and when $\lambda \to 0$, the probability that there are two distinct edges from b^* of weight smaller than $X - Y_i$ is of order $O(\lambda^2)$. Indeed, let $\mathcal{F}$ denote the existence of two distinct edges from b^* of weight smaller than X and let $\mathcal{F}_{i,j}$ denote the event that (a_i, b^*) and (a_j, b^*) both have weight smaller than X.

Then,

$$\mathbb{P}(\mathcal{F}) \le n^2\, \mathbb{E}_X(\max_{i,j} \mathbb{P}(\mathcal{F}_{i,j} \mid X)) = n^2\, \mathbb{E}((1 - e^{-\lambda X})^2) \le n^2\lambda^2\, \mathbb{E}(X^2), \tag{14.11}$$

and since $\mathbb{E}(X^2)$ is finite and independent of λ, this is $O(\lambda^2)$.

Note that Y and Y_i have the same distribution. They are both equal to the minimum weight of a matching of a random $(r-1)$ set of A into B. As a consequence, $\mathbb{E}(Y) = \mathbb{E}(Y_i) = \frac{1}{r}\sum_{j\in A_r} \mathbb{E}(Y_j)$. Since w_i is $EXP(\lambda)$ distributed, as $\lambda \to 0$, we have from (14.11) that

$$\begin{aligned} P(n,r) &= \lim_{\lambda\to 0}\left(\frac{1}{\lambda}\sum_{j\in A_r} \mathbb{P}(w_j < X - Y_j) + O(\lambda)\right) \\ &= \lim_{\lambda\to 0} \mathbb{E}\left(\frac{1}{\lambda}\sum_{j\in A_r}\left(1 - e^{-\lambda(X-Y_j)}\right)\right) = \sum_{j\in A_r} \mathbb{E}(X - Y_i) = r\,\mathbb{E}(X - Y). \end{aligned}$$

□

We now proceed to estimate $P(n,r)$. Fix r and assume that $b^* \notin B_{r-1}^*$. Suppose that M_r^* is obtained from M_{r-1}^* by finding an augmenting path $P = (a_r, \dots, a_\sigma, b_\tau)$ from a_r to $B \setminus B_{r-1}$ of minimum additional weight. We condition on (i) σ, (ii) the lengths of all edges other than $(a_\sigma, b_j), b_j \in B \setminus B_{r-1}$, and (iii) $\min\{w(a_\sigma, b_j) : b_j \in B \setminus B_{r-1}\}$. With this conditioning, $M_{r-1} = M_{r-1}^*$ will be fixed and so will $P' = (a_r, \dots, a_\sigma)$. We can now use the following fact: Let $X_1, X_2, \dots, X_M$ be independent exponential random variables of rates $\lambda_1, \lambda_2, \dots, \lambda_M$. Then the probability that X_i is the smallest of them is $\lambda_i/(\lambda_1 + \lambda_2 + \cdots + \lambda_M)$. Furthermore, the probability stays the same if we condition on the value of $\min\{X_1, X_2, \dots, X_M\}$. Thus,

$$\mathbb{P}(b^* \in B_r^* \mid b^* \notin B_{r-1}^*) = \frac{\lambda}{n - r + 1 + \lambda}.$$

Lemma 14.6

$$P(n,r) = \frac{1}{n} + \frac{1}{n-1} + \cdots + \frac{1}{n-r+1}. \tag{14.12}$$

Proof

$$\lim_{\lambda\to 0}\lambda^{-1}\,\mathbb{P}(b^*\in B_r^*)=\lim_{\lambda\to 0}\lambda^{-1}\left(1-\frac{n}{n+\lambda}\cdot\frac{n-1}{n-1+\lambda}\cdots\frac{n-r+1}{n-r+1+\lambda}\right)$$
$$=\frac{1}{n}+\frac{1}{n-1}+\cdots+\frac{1}{n-r+1}. \tag{14.13}$$

□

It follows from Lemmas 14.5 and 14.6 that

$$\mathbb{E}\,C_n=\sum_{r=1}^{n}\frac{1}{r}\sum_{i=1}^{r}\frac{1}{n-i+1},$$

which implies that

$$\mathbb{E}(C_{n+1}-C_n)=\frac{1}{(n+1)^2}. \tag{14.14}$$

$\mathbb{E}(C_1)=1$ and so (14.8) follows from (14.14). □

Exercises

14.3.1 Verify equation (14.13).
14.3.2 Verify equation (14.14).

Problems for Chapter 14

14.1 Suppose that the edges of the complete bipartite graph $K_{n,n}$ are given independent uniform $[0, 1]$ edge weights. Show that if $L_n^{(b)}$ is the length of the MST, then

$$\lim_{n\to\infty} \mathbb{E} L_n^{(b)} = 2\zeta(3).$$

14.2 Let $G = K_{\alpha n,\beta n}$ be the complete unbalanced bipartite graph with bipartition sizes $\alpha n, \beta n$. Suppose that the edges of G are given independent uniform $[0, 1]$ edge weights. Show that if $L_n^{(b)}$ is the length of the MST, then

$$\lim_{n\to\infty} \mathbb{E} L_n^{(b)} = \gamma + \frac{1}{\gamma} + \sum_{i_1\geq 1, i_2\geq 1} \frac{(i_1 + i_2 - 1)!}{i_1! i_2!} \frac{\gamma^{i_1} i_1^{i_2-1} i_2^{i_1-1}}{(i_1 + \gamma i_2)^{i_1+i_2}},$$

where $\gamma = \alpha/\beta$.

14.3 Tighten Theorem 14.1 and prove that

$$\mathbb{E} L_n = \zeta(3) + O\left(\frac{1}{n}\right).$$

14.4 Suppose that the edges of K_n are given independent uniform $[0, 1]$ edge weights. Let Z_k denote the minimum total edge cost of the union of k edge-disjoint spanning trees. Show that $\lim_{k\to\infty} Z_k/k^2 = 1$.

14.5 Show that if the edges of the complete bipartite graph $K_{n,n}$ are given i.i.d. costs, then the minimum cost perfect matching is uniformly random among all $n!$ perfect matchings.

14.6 Show that if $\omega \to \infty$, then w.h.p. no edge in the minimum cost perfect matching has cost more than $\frac{1}{\omega \log n}$.

14.7 Show that a random permutation π gives rise to a digraph $D_\pi = ([n], \{(i, \pi(i)) : i \in [n]\})$ that w.h.p. consists of $O(\log n)$ vertex disjoint cycles that cover $[n]$.

14.8 Consider the Asymmetric Traveling Salesperson Problem (ATSP) where the costs $C(i, j)$ are independent uniform $[0, 1]$. Use the claimed results of the previous two problems to show that Karp's patching algorithm finds a tour that is within $(1 + o(1))$ of minimal cost w.h.p.

The ATSP asks for the minimum total cost of a directed Hamilton cycle through the complete digraph $\vec{K}_n$.

Karp's algorithm starts by solving the assignment problem with costs $C(i, j)$. It interprets the perfect matching as the union of disjoint cycles in $\vec{K}_n$ and then patches them together cheaply.

Given cycles C_1, C_2 and edges $e_i = (x_i, y_i) \in C_i, i = 1, 2$, a patch removes e_1, e_2 and replaces them with (x_2, y_1) plus (x_1, y_2) creating a single cycle.

14.9 Suppose that the edges of $G_{n,p}$, where $0 < p \leq 1$ is a constant, are given exponentially distributed weights with rate 1. Show that if X_{ij} is the shortest distance from i to j, then

1 for any fixed i, j,

$$\mathbb{P}\left(\left|\frac{X_{ij}}{\log n/n} - \frac{1}{p}\right| \geq \varepsilon\right) \to 0;$$

2

$$\mathbb{P}\left(\left|\frac{\max_j X_{ij}}{\log n/n} - \frac{2}{p}\right| \geq \varepsilon\right) \to 0.$$

14.10 The *quadratic assignment problem* is to

Minimize

$$Z = \sum_{i,j,p,q=1}^{n} a_{ijpq} x_{ip} x_{jq}$$

Subject to

$$\sum_{i=1}^{n} x_{ip} = 1 \qquad p = 1, 2, \ldots, n,$$

$$\sum_{p=1}^{n} x_{ip} = 1 \qquad i = 1, 2, \ldots, n,$$

$$x_{ip} = 0/1.$$

Suppose now that the a_{ijpq} are independent uniform $[0, 1]$ random variables. Show that w.h.p. $Z_{\min} \sim Z_{\max}$, where $Z_{\min}$ (resp. $Z_{\max}$) denotes the minimum (resp. maximum) value of Z, subject to the assignment constraints.

14.11 The *0/1 knapsack problem* is to

Maximize

$$Z = \sum_{i=1}^{n} a_i x_i$$

Subject to

$$\sum_{i=1}^{n} b_i x_i \leq L,$$

$$x_i = 0/1 \quad \text{for } i = 1, 2, \ldots, n.$$

Suppose that the (a_i, b_i) are chosen independently and uniformly from $[0, 1]^2$ and that $L = \alpha n$. Show that w.h.p. the maximum value of Z, $Z_{\max}$, satisfies

$$Z_{\max} \sim \begin{cases} \frac{\alpha^{1/2} n}{2} & \alpha \leq \frac{1}{4}, \\ \frac{(8\alpha - 8\alpha^2 - 1)n}{2} & \frac{1}{4} \leq \alpha \leq \frac{1}{2}, \\ \frac{n}{2} & \alpha \geq \frac{1}{2}. \end{cases}$$

14.12 Suppose that $X_1, X_2, \ldots, X_n$ are points chosen independently and uniformly at random from $[0, 1]^2$. Let Z_n denote the total Euclidean length of the shortest tour (Hamilton cycle) through each point. Show that there exist constants c_1, c_2 such that $c_1 n^{1/2} \leq Z_n \leq c_2 n^{1/2}$ w.h.p.

References

[1] M. Abdullah and N. Fountoulakis, A phase transition on the evolution of bootstrap percolation processes on preferential attachment graphs, *Random Structures and Algorithms*, **52** (2018) 379–418.

[2] W. Aiello, W. Bonato, C. Cooper, J. Janssen and P. Prałat, A spatial web graph model with local influence regions, *Internet Mathematics*, **5** (2009) 175–196.

[3] M. Ajtai, J. Komlós and E. Szemerédi, The longest path in a random graph, *Combinatorica*, **1** (1981) 1–12.

[4] M. Ajtai, J. Komlós and E. Szemerédi, The first occurrence of Hamilton cycles in random graphs, *Annals of Discrete Mathematics*, **27** (1985) 173–178.

[5] D. Aldous, Asymptotics in the random assignment problem, *Probability Theory and Related Fields*, **93** (1992) 507–534.

[6] D. Aldous, The $\zeta(2)$ limit in the random assignment problem, *Random Structures and Algorithms*, **4** (2001) 381–418.

[7] N. Alon, A note on network reliability. In *Discrete Probability and Algorithms* (Minneapolis, MN, 1993), The IMA Volumes in Mathematics and Its Applications, Vol. **72**, Springer (1995) 11–14.

[8] N. Alon and Z. Füredi, Spanning subgraphs of random graphs, *Graphs and Combinatorics*, **8** (1992) 91–94.

[9] L. Babai, P. Erdős and S. M. Selkow, Random graph isomorphism, *SIAM Journal on Computing*, **9** (1980) 628–635.

[10] D. Bal, P. Bennett, A. Dudek and A. M. Frieze, The t-tone chromatic number of random graphs, *Graphs and Combinatorics*, **30** (2014) 1073–1086.

[11] F. Ball, D. Mollison and G. Scalia-Tomba, Epidemics with two levels of mixing, *Annals of Applied Probability*, **7** (1997) 46–89.

[12] J. Balogh, B. Bollobás, M. Krivelevich, T. Müeller and M. Walters, Hamilton cycles in random geometric graphs, *Annals of Applied Probability*, **21** (2011) 1053–1072.

[13] L. Barabási and R. Albert, Emergence of scaling in random networks, *Science*, **286** (1999) 509–512.

[14] A. D. Barbour, M. Karoński and A. Ruciński, A central limit theorem for decomposable random variables with applications to random graphs, *Journal of Combinatorial Theory B*, **47** (1989) 125–145.

[15] A. D. Barbour and G. Reinert, Small worlds, *Random Structures and Algorithms*, **19** (2001) 54–74.

[16] A. D. Barbour and G. Reinert, Discrete smallworld networks, *Electronic Journal on Probability*, **11** (2006) 1234–1283.

[17] S. N. Bernstein, *Theory of Probability*, (in Russian), Gostekhizdat (1927).

[18] B. Bollobás, A probabilistic proof of an asymptotic formula for the number of labelled graphs, *European Journal on Combinatorics*, **1** (1980) 311–316.

[19] B. Bollobás, Random graphs. In *Combinatorics, Proceedings, Swansea*, London Mathematical Society Lecture Note Series, Vol. 52, Cambridge University Press (1981) 80–102.

[20] B. Bollobás, The evolution of random graphs, *Transactions of the American Mathematical Society*, **286** (1984) 257–274.

[21] B. Bollobás, *Random Graphs*, First edition, Academic Press (1985); Second edition, Cambridge University Press (2001).

[22] B. Bollobás, The chromatic number of random graphs, *Combinatorica*, **8** (1988) 49–56.

[23] B. Bollobás and F. Chung, The diameter of a cycle plus a random matching, *SIAM Journal on Discrete Mathematics*, **1** (1988) 328–333.

[24] B. Bollobás and P. Erdős, Cliques in random graphs, *Mathematical Proceedings of the Cambridge Philosophical Society*, **80** (1976) 419–427.

[25] B. Bollobás and A. Frieze, On matchings and Hamiltonian cycles in random graphs, *Annals of Discrete Mathematics*, **28** (1985) 23–46.

[26] B. Bollobás and A. Frieze, Spanning maximal planar subgraphs of random graphs, *Random Structures and Algorithms*, **2** (1991) 225–231.

[27] B. Bollobás and O. Riordan, The diameter of a scale free random graph, *Combinatorica*, **24** (2004) 5–34.

[28] B. Bollobás, O. Riordan, J. Spencer and G. Tusnády, The degree sequence of a scale-free random graph process, *Random Structures and Algorithms*, **18** (2001) 279–290.

[29] B. Bollobás and A. Thomason, Threshold functions, *Combinatorica*, **7** (1987) 35–38.

[30] F. Chung and L. Lu, Connected components in random graphs with given expected degree sequence, *Annals of Combinatorics*, **6** (2002) 125–145.

[31] F. Chung and L. Lu, The volume of the giant component of a random graph with given expected degrees, *SIAM Journal on Discrete Mathematics*, **20** (2006) 395–411.

[32] F. Chung and L. Lu, *Complex Graphs and Networks*, American Mathematical Society (2006).

[33] R. Cont and E. Tanimura, Small-world graphs: characterizations and alternative constructions, *Advances in Applied Probability*, **20** (2008) 939–965.

[34] C. Cooper, Pancyclic Hamilton cycles in random graphs, *Discrete Mathematics*, **91** (1991) 141–148.

[35] C. Cooper, 1-pancyclic Hamilton cycles in random graphs, *Random Structures and Algorithms*, **3** (1992) 277–287.

[36] C. Cooper and A. M. Frieze, Pancyclic random graphs. In *Random Graphs*, M. Karonski, J. Javorski, and A. Rucinski (eds). John Wiley and Sons (1990) 29–39.

[37] C. Cooper, A. Frieze and P. Prałat. Some typical properties of the spatial preferred attachment model, *Internet Mathematics*, **10** (2014) 27–47.

[38] J. Díaz, D. Mitsche and X. Pérez-Giménez, Sharp threshold for hamiltonicity of random geometric graphs, *SIAM Journal on Discrete Mathematics*, **21** (2007) 57–65.

[39] A. Dudek, D. Mitsche and P. Prałat, The set chromatic number of random graphs, *Discrete Applied Mathematics*, **215** (2016) 61–70.

[40] A. Dudek and P. Prałat, An alternative proof of the linearity of the size-Ramsey number of paths, *Combinatorics, Probability and Computing*, **24** (2015) 551–555.

[41] R. Durrett, *Random Graph Dynamics*, Cambridge University Press (2010).

[42] P. Erdős and A. Rényi, On random graphs I, *Publicationes Mathematicae Debrecen*, **6** (1959) 290–297.

[43] P. Erdős and A. Rényi, On the evolution of random graphs, *Publication of the Mathematical Institute of the Hungarian Academy of Sciences*, **5** (1960) 17–61.

[44] P. Erdős and A. Rényi, On the strength of connectedness of a random graph, *Acta Mathematica Academiae Scientiarum Hungaricae*, **8** (1961) 261–267.

[45] P. Erdős and A. Rényi, On random matrices, *Publication of the Mathematical Institute of the Hungarian Academy of Sciences*, **8** (1964) 455–461.

[46] P. Erdős and A. Rényi, On the existence of a factor of degree one of a connected random graph, *Acta mathematica Academiae Scientiarum Hungaricae*, **17** (1966) 359–368.

[47] M. Faloutsos, P. Faloutsos and C. Faloutsos, On power-law relationships of the internet topology, *ACM SIGCOMM* (1999).

[48] J. A. Fill, E. R. Scheinerman and K. B. Singer-Cohen, Random intersection graphs when $m = \omega(n)$: An equivalence theorem relating the evolution the evolution of the $G(n, m, p)$ and $G(n, p)$ models, *Random Structures and Algorithms*, **16** (2000) 156–176.

[49] A. Flaxman, A. M. Frieze and T. Fenner, High degree vertices and eigenvalues in the preferential attachment graph, *Internet Mathematics*, **2** (2005) 1–20.

[50] A. M. Frieze, On the value of a random minimum spanning tree problem, *Discrete Applied Mathematics*, **10** (1985) 47–56.

[51] A. M. Frieze, On the independence number of random graphs, *Discrete Mathematics*, **81** (1990) 171–176.

[52] A. M. Frieze and M. Karoński, *Introduction to Random Graphs*, Cambridge University Press (2016).

[53] A. M. Frieze and C. J. H. McDiarmid, On random minimum length spanning trees, *Combinatorica*, **9** (1989) 363–374.

[54] E. N. Gilbert, Random graphs, *Annals of Mathematical Statistics*, **30** (1959) 1141–1144.

[55] G. Grimmett and C. McDiarmid, On colouring random graphs, *Mathematical Proceedings of the Cambridge Philosophical Society*, **77** (1975) 313–324.

[56] L. Gu and L. H. Huang, The clustering coefficient and the diameter of small-world networks, *Acta Mathematica Sinica, English Series*, **29** (2013) 199–208.

[57] P. Gupta and P. R. Kumar, Critical power for asymptotic connectivity in wireless networks. In *Stochastic Analysis, Control, Optimization and Applications: A Volume in Honor of W. H. Fleming*, W. M. McEneany, G. Yin and Q. Zhang (eds). Birkhauser (1998) 547–566.

[58] E. Győri, B. Rothschild and A. Ruciński, Every graph is contained in a sparest possible balanced graph, *Mathematical Proceedings of the Cambridge Philosophical Society*, **98** (1985) 397–401.

[59] W. Hoeffding, Probability inequalities for sums of bounded random variables, *Journal of the American Statistical Association*, **58** (1963) 13–30.

[60] R. van der Hofstad, *Random Graphs and Complex Networks*, Vol. 1, Cambridge University Press, 2017.

[61] J. Janssen, P. Prałat and R. Wilson, Geometric graph properties of the spatial preferred attachment model, *Advances in Applied Mathematics*, **50** (2013) 243–267.

[62] S. Janson, The minimal spanning tree in a complete graph and a functional limit theorem for trees in a random graph, *Random Structures and Algorithms*, **7** (1995) 337–355.

[63] S. Janson, One, two and three times $\log n/n$ for paths in a complete graph with random weights, *Combinatorics, Probability and Computing*, **8** (1999) 347–361.

[64] S. Janson, Asymptotic degree distribution in random recursive trees, *Random Structures and Algorithms*, **26** (2005) 69–83.

[65] S. Janson, Plane recursive trees, Stirling permutations and an urn model. In *Fifth Colloquium on Mathematics and Computer Science*, Discrete Math. Theor. Comput. Sci. Proc., AI, Discrete *Mathematics and Theoretical Computer Science*, (2008) 541–547.

[66] S. Janson, T. Łuczak and A. Ruciński, *Random Graphs*, John Wiley and Sons (2000).

[67] D. Karger and C. Stein, A new approach to the minimum cut problem, *Journal of the ACM*, **43** (1996) 601–640.

[68] M. Karoński, E. Scheinerman and K. Singer-Cohen, On random intersection graphs: the subgraph problem, *Combinatorics, Probability and Computing*, **8** (1999) 131–159.

[69] M. Karoński and A. Ruciński, On the number of strictly balanced subgraphs of a random graph, In *Graph Theory, Proc. Łagów, 1981*, Lecture Notes in Mathematics 1018, Springer (1983) 79–83.

[70] M. Karoński and A. Ruciński, Problem 4, In *Graphs and other Combinatorial Topics, Proceedings, Third Czech. Symposium on Graph Theory, Prague* (1983).

[71] M. Karoński and A. Ruciński, Poisson convergence and semi-induced properties of random graphs, *Mathematical Proceedings of the Cambridge Philosophical Society*, **101** (1987) 291–300.

[72] J. Kleinberg, The small-world phenomenon: An algorithmic perspective, *Proceedings of the 32nd ACM Symposium on Theory of Computing* (2000) 163–170.

[73] J. Komlós and E. Szemerédi, Limit distributions for the existence of Hamilton circuits in a random graph, *Discrete Mathematics*, **43** (1983) 55–63.

[74] W. Kordecki, Normal approximation and isolated vertices in random graphs, In *Random Graphs '87*, M. Karoński, J. Jaworski and A. Ruciński (eds). John Wiley and Sons (1990) 131–139.

[75] I. N. Kovalenko, Theory of random graphs, *Cybernetics and Systems Analysis*, **7** (1971) 575–579.

[76] M. Krivelevich, C. Lee and B. Sudakov, Long paths and cycles in random subgraphs of graphs with large minimum degree, *Random Structures and Algorithms*, **46** (2015) 320–345.

[77] M. Krivelevich and B. Sudakov, The phase transition in random graphs – a simple proof, *Random Structures and Algorithms*, **43** (2013) 131–138.

[78] S. Linusson and J. Wästlund, A proof of Parisi's conjecture on the random assignment problem, *Probability Theory and Related Fields*, **128** (2004) 419–440.

[79] T. Łuczak, On the equivalence of two basic models of random graphs, In *Proceedings of Random Graphs '87*, M. Karonski, J. Jaworski, and A. Rucinski (eds). John Wiley and Sons (1990) 151–158.

[80] T. Łuczak, Component behaviour near the critical point, *Random Structures and Algorithms*, **1** (1990) 287–310.

[81] T. Łuczak, On the number of sparse connected graphs, *Random Structures and Algorithms*, **1** (1990) 171–174.

[82] T. Łuczak, Cycles in a random graph near the critical point, *Random Structures and Algorithms*, **2** (1991) 421–440.

[83] T. Łuczak, The chromatic number of random graphs, *Combinatorica*, **11** (1991) 45–54.

[84] E. Marczewski, Sur deux propriétés des classes d'ensembles, *Fundamenta Mathematicae*, **33** (1945) 303–307.

[85] D. Matula, The largest clique size in a random graph, *Technical Report, Department of Computer Science, Southern Methodist University*, (1976).

[86] D. Matula, Expose-and-merge exploration and the chromatic number of a random graph, *Combinatorica*, **7** (1987) 275–284.

[87] C. McDiarmid, Random channel assignment in the plane, *Random Structures and Algorithms*, **22** (2003) 187–212.

[88] C. McDiarmid and B. Reed, Colouring proximity graphs in the plane, *Discrete Mathematics*, **199** (1999) 123–137.

[89] S. Milgram, The Small World Problem, *Psychology Today*, **2** (1967) 60–67.

[90] M. Molloy and B. Reed, The Size of the Largest Component of a random graph on a fixed degree sequence, *Combinatorics, Probability and Computing*, **7** (1998) 295–306.

[91] T. Müller, X. Pérez and N. Wormald, Disjoint Hamilton cycles in the random geometric graph, *Journal of Graph Theory*, **68** (2011) 299–322.

[92] T. Müller, Private communication.

[93] A. Nachmias and Y. Peres, The critical random graph, with martingales, *Israel Journal of Mathematics*, **176** (2010) 29–41.

[94] C. Nair, B. Prabhakar and M. Sharma, Proofs of the Parisi and Coppersmith–Sorkin random assignment conjectures, *Random Structures and Algorithms*, **27** (2005) 413–444.

[95] M. Newman, *Networks*, Second Edition, Oxford University Press, 2018.

[96] M. E. J. Newman and D. J. Watts, Scaling and percolation in the small-world network model, *Physical Review E*, **60** (1999) 7332–7342.

[97] E. Palmer and J. Spencer, Hitting time for k edge disjoint spanning trees in a random graph, *Periodica Mathematica Hungarica*, **91** (1995) 151–156.

[98] G. Parisi, A conjecture on Random Bipartite Matching, Physics e-Print archive (1998).

[99] M. Penrose, On k-connectivity for a geometric random graph, *Random Structures and Algorithms*, **15** (1999) 145–164.

[100] M. Penrose, *Random Geometric Graphs*, Oxford University Press, 2003.

[101] L. Pósa, Hamiltonian circuits in random graphs, *Discrete Mathematics*, **14** (1976) 359–364.

[102] A. Ruciński, When a small subgraphs of a random graph are normally distributed, *Probability Theory and Related Fields*, **78** (1988) 1–10.

[103] A. Ruciński and A. Vince, Strongly balanced graphs and random graphs, *Journal of Graph Theory*, **10** (1986) 251–264.

[104] K. Rybarczyk, Equivalence of the random intersection graph and $G(n, p)$, *Random Structures and Algorithms*, **38** (2011) 205–234.

[105] K. Rybarczyk and D. Stark, Poisson approximation of the number of cliques in a random intersection graph, *Journal of Applied Probability*, **47** (2010) 826–840.

[106] B. Sudakov and V. Vu, Local resilience of graphs, *Random Structures and Algorithms*, **33** (2008) 409–433.

[107] A. Scott, On the concentration of the chromatic number of random graphs (2008), arxiv 0806.0178

[108] J. Spencer with L. Florescu, *Asymptotia*, American Mathemetical Society, 2010.

[109] J. M. Steele, On Frieze's $\zeta(3)$ limit for lengths of minimal spanning trees, *Discrete Applied Mathematics*, **18** (1987) 99–103.

[110] P. Turán, On an extremal problem in graph theory (in Hungarian), *Matematikai és Fizikai Lapok*, **48** (1941) 436–452.

[111] W. F. de la Vega, Long paths in random graphs, *Studia Scientiarum Mathematicarum Hungarica*, **14** (1979) 335–340.

[112] J. Wästlund, An easy proof of the $\zeta(2)$ limit in the random assignment problem, *Electronic Communications in Probability*, **14** (2009) 261–269.

[113] D. J. Watts and S. H. Strogatz, Collective dynamics of "small-world" networks, *Nature*, **393** (1998) 440–442.

[114] D. West, *Introduction to Graph Theory*, Second edition, Prentice Hall 2001.

[115] N. Wormald, Models of random regular graphs, In *Surveys in Combinatorics 1999*, J. Lamb and D. Preece (eds). London Mathematical Society Lecture Note Series (1999) 239–298, Cambridge University Press.

Author Index

Subject Index

Printed in the United States
by Baker & Taylor Publisher Services